Lecture Notes in Physics

Volume 883

T0155916

For further volumes:
www.springer.com/series/5304

The Lecture Notes in Physics

The series Lecture Notes in Physics (LNP), founded in 1969, reports new developments in physics research and teaching—quickly and informally, but with a high quality and the explicit aim to summarize and communicate current knowledge in an accessible way. Books published in this series are conceived as bridging material between advanced graduate textbooks and the forefront of research and to serve three purposes:

- to be a compact and modern up-to-date source of reference on a well-defined topic
- to serve as an accessible introduction to the field to postgraduate students and nonspecialist researchers from related areas
- to be a source of advanced teaching material for specialized seminars, courses and schools

Both monographs and multi-author volumes will be considered for publication. Edited volumes should, however, consist of a very limited number of contributions only. Proceedings will not be considered for LNP.

Volumes published in LNP are disseminated both in print and in electronic formats, the electronic archive being available at springerlink.com. The series content is indexed, abstracted and referenced by many abstracting and information services, bibliographic networks, subscription agencies, library networks, and consortia.

Proposals should be sent to a member of the Editorial Board, or directly to the managing editor at Springer:

Christian Caron
Springer Heidelberg
Physics Editorial Department I
Tiergartenstrasse 17
69121 Heidelberg/Germany
christian.caron@springer.com

Johannes M. Henn · Jan C. Plefka

Scattering Amplitudes
in Gauge Theories

 Springer

Dr. Johannes M. Henn
School of Natural Sciences
Institute for Advanced Study
Princeton, NJ, USA

Prof. Dr. Jan C. Plefka
Institut für Physik
Math.-Naturw. Fakultät
Humboldt Universität zu Berlin
Berlin, Germany

ISSN 0075-8450   ISSN 1616-6361 (electronic)
Lecture Notes in Physics
ISBN 978-3-642-54021-9   ISBN 978-3-642-54022-6 (eBook)
DOI 10.1007/978-3-642-54022-6
Springer Heidelberg New York Dordrecht London

Library of Congress Control Number: 2014932295

Printed on acid-free paper

Springer is part of Springer Science+Business Media (www.springer.com)

To Mieke
JP

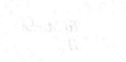

Preface

In our current understanding of subatomic particle physics, recently successfully tested with the discovery of the Higgs boson, processes involving elementary particles are described by quantum field theory. Scattering amplitudes are the central objects in perturbative quantum field theory as they link the theoretical description to experimental predictions. To a large degree the historical development of quantum field theory was driven by the need to compute scattering amplitudes. They constitute the integral building block for the construction of scattering cross sections determining the probabilities for scattering processes to occur at particle colliders. Non-abelian gauge field or Yang-Mills theories represent the backbone of high energy physics, as they provide the theoretical framework to describe the interactions of elementary particles in the standard model. The principles for computing scattering amplitudes in gauge theories have been settled since the mid 1970s: once the Lagrangian and its gauge fixing term has been constructed, one reads off the Feynman rules for the scalar, fermion and gauge fields. The scattering amplitude is then given by the sum of all contributing diagrams built from the Feynman rules with the subsequent integration over the internal loop momenta. Divergences in these integrals require regularization, and the ultraviolet divergences lead to a renormalization of the theory. So from a conceptual viewpoint one might consider this chapter of quantum field theory as complete.

However, it turns out that beyond the simplest examples the complexity of the Feynman diagrammatic computation quickly gets out of hand. Already at tree-level such a computation can become enormous if one is dealing with gauge interactions. For example, the number of Feynman diagrams contributing to a gluon tree-level amplitude $g + g \to n \cdot g$ grows factorially with n [1]. On the other hand, the final answer can be obtained in closed analytic form, and, when expressed in convenient variables, is remarkably simple. There are two complementary ways of understanding the simplicity of the final answer. A reason for the complexity of the Feynman diagram calculation is that individual Feynman diagrams are gauge variant and involve off-shell intermediate states in internal propagators. The amplitude, on the other hand, is gauge invariant and only knows about on-shell degrees of freedom. Hence, in going from Feynman diagrams to an amplitude the unphysical degrees of freedom cancel. On-shell approaches that focus on the analytic structure of the final result allow to circumvent these unnecessary complications. Another reason for the simplicity of the final answer has to do with symmetry properties and is more surprising: it turns out that besides the obvious symmetries of the Lagrangian, gluon amplitudes have additional, hidden symmetries, that constrain their form. Choosing appropriate variables makes the action of these symmetries transparent and thereby also simplifies the expressions. For these reasons, both analyticity and symmetry properties are two topics that will play an important role in these lecture notes.

In the past decade tremendous progress was made both in our understanding of the structure of scattering amplitudes and in the ability of computing the latter in gauge theories at high multiplicities (number of external legs) and beyond the one-loop order. These advances provided the field with a new set of tools that go beyond the textbook approaches. These methods build on using a color decomposition of the gauge theory amplitudes and on expressing them in a spinor helicity basis particularly suited for massless particles. Thinking about the analytic structure of tree-level amplitudes leads to novel on-shell recursion relations. They allow the analytic construction of tree-level amplitudes from atomistic three-point ones. At loop level unitarity-based techniques, combined with the knowledge of an integral basis for one-loop Feynman integrals, may be used to construct loop amplitudes from tree-level amplitudes. In summary, all amplitudes follow from the on-shell three-point vertices, and no reference to the complicated form of the Lagrangian, gauge fixing terms and ghosts is necessary.

Our main focus is on methods based on general principles such as analyticity and symmetries. The progress for generic gauge theories was often led by studies in the maximally supersymmetric gauge theory in four dimensions, $\mathcal{N} = 4$ super Yang-Mills. It may be thought of as an idealized version of QCD and provides an ideal laboratory for the development and testing of new approaches. In fact, gluon tree-level amplitudes in both models are identical, but they can be easily solved for within the $\mathcal{N} = 4$ super Yang-Mills framework, in part thanks to fascinating hidden symmetries. Discussing this theory in these lecture notes also allows us to have many examples where scattering amplitudes can be computed analytically, and we hope that this will help to illustrate the general properties.

These modern helicity amplitude or on-shell methods are so far not treated in depth in standard textbooks on quantum field theory. Yet they are of increasing importance in phenomenological applications, which face the demand of high precision predictions for high multiplicity scattering events at the Large Hadron Collider, as well as in foundational studies in quantum gauge field theory towards the structure, possible symmetries and construction of the perturbative S-matrix. In fact these developments have led to a cross-fertilization of high-energy phenomenology and formal theory in recent years. This has made the field of scattering amplitudes a very active and fascinating area of research attracting many new researchers and students. The goal of these lecture notes in physics is to serve these communities in bridging the gap from a textbook knowledge of quantum field theory to a working expertise for doing research in the domain of scattering amplitudes. Our aim was to keep these notes compact so that they are suitable for an advanced graduate lecture course. In order to make them more accessible for students, we have included a series of exercises, together with their solutions. Due to our wish to keep the volume of these notes amenable to a lecture course, a certain selection of topics, certainly biased by personal tastes, had to be made. At the end of each chapter we therefore give an account of related developments in the field not discussed in the lecture notes and give references to all relevant works.

The level of the text aims at advanced graduate students who have already followed an introductory course on quantum field theory, typically leading to QED scattering processes and a few simple one-loop computations therein. The basics of non-abelian quantum field theories are briefly reviewed in chapter one of this book. Here also the basic tools of the modern helicity amplitude approach are provided and some simple tree-level diagrams are computed in the conventional way, using color-ordered Feynman rules. Chapter two introduces the on-shell recursion and discusses universal factorization properties for color ordered amplitudes. After a discussion of Poincaré and conformal symmetry, the $\mathcal{N} = 4$ super Yang-Mills theory is introduced, along with its super-amplitude formalism and a supersymmetric on-shell recursion relation allowing for an exact solution. In chapter three the loop-level structure of amplitudes is reviewed. Here the reduction of one-loop Feynman integrals to a basis of scalar integrals is discussed. The idea of (generalized) unitarity in constructing one-loop amplitudes from tree-level data by putting various internal legs on-shell is reviewed and a number of concrete examples are computed in detail. In the second part of chapter three we give an introduction to the evaluation of Feynman integrals at one and higher loop order. After reviewing their definition and useful parameter representations, such as the Feynman and Mellin representations, we give an introduction to the integration by parts techniques in conjunction with differential equations. The final chapter is devoted to advanced topics mostly within maximally supersymmetric gauge theory: recursion relations for loop integrands, the duality of scattering amplitudes to Wilson loops with light-like contours and correlation functions of local operators. Finally, the hidden dual conformal and Yangian symmetries of $\mathcal{N} = 4$ super Yang-Mills amplitudes are discussed pointing towards a fascinating hidden integrability of this four-dimensional gauge theory.

Some elements of modern on-shell methods are also discussed in the recent quantum field theory textbooks of Srednicki [2] and Zee [3]. Introductory and in-depth reviews exist on the various topics presented in these lecture notes. These are [1, 4–6] on the topics of chapters one and two. The integral reduction and unitarity methods are pedagogically reviewed in [7, 8] from a phenomenological viewpoint. The comprehensive textbook of Smirnov [9, 10] discusses techniques for the computation of Feynman integrals in great detail. As for more formal aspects related to $\mathcal{N} = 4$ super Yang-Mills, symmetries and dualities, there exists a special issue of J. Phys. A [11–24], as well as the very recent comprehensive review [25]. These references have overlaps with some of the material presented in these notes, but also discuss supergravity amplitudes, dualities of gauge and gravity amplitudes, amplitudes in twistor space, and the Grassmannian approach to superamplitudes. This is certainly not a complete list but reflects the texts from which we have learned a lot ourselves.

This manuscript grew out of lecture notes of a physics master's level course taught at Humboldt University Berlin in the summer terms of 2011 and 2013, as well as a series of shorter lecture series delivered at research schools in Atrani (Italy), Dubna (Russia), Durham (UK), Copenhagen (Denmark), Parma (Italy), Waterloo (Canada) and Wolfersdorf (Germany) in the years 2010 to 2013.

We would like to thank the students and participants of these lectures and schools for questions and comments on the manuscript, in particular L. Bianchi, M. Heinze, D. Müller and H. Münkler. We thank H. Stephan for help with typesetting the graphs and diagrams. We would like to thank our colleagues V. Smirnov, P. Uwer and G. Yang for important comments on the manuscript. Finally, we would like to thank N. Arkani-Hamed, S. Badger, N. Beisert, Z. Bern, B. Biedermann, R. Boels, L. Ferro, S. Caron-Huot, L. Dixon, J. Drummond, H. Elvang, H. Johanssen, G. Korchemsky, D. Kosower, T. Lukowski, T. McLoughlin, S. Moch, T. Schuster, E. Sokatchev, M. Staudacher, and P. Uwer for important discussions or pleasant collaborations on topics related to these lecture notes.

References

1. M.L. Mangano, S.J. Parke, Multi-parton amplitudes in gauge theories. Phys. Rep. **200**, 301–367 (1991). arXiv:hep-th/0509223
2. M. Srednicki, Quantum Field Theory (Cambridge University Press, Cambridge, 2007)
3. A. Zee, Quantum Field Theory in a Nutshell (Princeton University Press, Princeton, 2003)
4. L.J. Dixon, Calculating scattering amplitudes efficiently (1996). arXiv:hep-ph/9601359
5. M.E. Peskin, Simplifying multi-Jet QCD computation (2011). arXiv:1101.2414
6. L.J. Dixon, A brief introduction to modern amplitude methods (2013). arXiv:1310.5353
7. Z. Bern, L.J. Dixon, D.A. Kosower, On-shell methods in perturbative QCD. Ann. Phys. **322**, 1587–1634 (2007). arXiv:0704.2798
8. R.K. Ellis, Z. Kunszt, K. Melnikov, G. Zanderighi, One-loop calculations in quantum field theory: from Feynman diagrams to unitarity cuts. Phys. Rep. **518**, 141–250 (2012). arXiv:1105.4319
9. V.A. Smirnov, Evaluating Feynman integrals. Tracts Mod. Phys. **211**, 1 (2004)
10. V.A. Smirnov, Analytic tools for Feynman integrals. Tracts Mod. Phys. **250**, 1 (2012)
11. M.S. Radu Roiban, A. Volovich, Scattering amplitudes in gauge theories: progress and outlook. J. Phys. A **44**, 450301 (2011)
12. L.J. Dixon, Scattering amplitudes: the most perfect microscopic structures in the universe. J. Phys. A **44**, 454001 (2011). arXiv:1105.0771
13. A. Brandhuber, B. Spence, G. Travaglini, Tree-level formalism. J. Phys. A **44**, 454002 (2011). arXiv:1103.3477
14. Z. Bern, Y.-t. Huang, Basics of generalized unitarity. J. Phys. A **44**, 454003 (2011). arXiv:1103.1869
15. J.J.M. Carrasco, H. Johansson, Generic multiloop methods and application to $N = 4$ super-Yang-Mills. J. Phys. A **44**, 454004 (2011). arXiv:1103.3298
16. H. Ita, Susy theories and QCD: numerical approaches. J. Phys. A **44**, 454005 (2011). arXiv:1109.6527
17. R. Britto, Loop amplitudes in gauge theories: modern analytic approaches. J. Phys. A **44**, 454006 (2011). arXiv:1012.4493
18. R.M. Schabinger, One-loop $N = 4$ super Yang-Mills scattering amplitudes in d dimensions, relation to open strings and polygonal Wilson loops. J. Phys. A **44**, 454007 (2011). arXiv:1104.3873

19. T. Adamo, M. Bullimore, L. Mason, D. Skinner, Scattering amplitudes and Wilson loops in twistor space. J. Phys. A **44**, 454008 (2011). arXiv:1104.2890

20. H. Elvang, D.Z. Freedman, M. Kiermaier, SUSY ward identities, superamplitudes, and counterterms. J. Phys. A **44**, 454009 (2011). arXiv:1012.3401

21. J.M. Drummond, Tree-level amplitudes and dual superconformal symmetry. J. Phys. A **44**, 454010 (2011). arXiv:1107.4544

22. J.M. Henn, Dual conformal symmetry at loop level: massive regularization. J. Phys. A **44**, 454011 (2011). arXiv:1103.1016

23. T. Bargheer, N. Beisert, F. Loebbert, Exact superconformal and Yangian symmetry of scattering amplitudes. J. Phys. A **44**, 454012 (2011). arXiv:1104.0700

24. J. Bartels, L.N. Lipatov, A. Prygarin, Integrable spin chains and scattering amplitudes. J. Phys. A **44**, 454013 (2011). arXiv:1104.0816

25. H. Elvang, Y.-t. Huang, Scattering amplitudes (2013). arXiv:1308.1697

Princeton, NJ, USA and Berlin, Germany   Johannes H. Henn
November 2013   Jan C. Plefka

Contents

Chapter 1
Introduction and Basics

In this chapter we briefly discuss the basics of quantum (gauge) field theory. The intention is to establish our notation and to recapitulate the essentials of gauge field theory to be used in subsequent chapters. This chapter provides a brief introduction into the rich subject of quantum gauge field theory, which is also reviewed in detail in standard text books on the subject.

1.1 Lorentz and Poincaré Group, Algebra and Representations

The fundamental symmetry group of all relativistic quantum field theories in four dimensions is the Lorentz group $SO(1, 3)$, which together with the translation group forms the Poincaré group. A Lorentz transformation is a linear homogeneous coordinate transformation which leaves the relativistic length x^2 invariant,

$$x'^\mu = \Lambda^\mu{}_\nu x^\nu, \quad \text{with } x'^2 = x^2 = \eta_{\mu\nu} x^\mu x^\nu, \tag{1.1}$$

where $\eta_{\mu\nu} = \text{diag}(+, -, -, -)$ denotes the Minkowski metric. This implies the defining condition

$$\eta_{\mu\nu} \Lambda^\mu{}_\rho \Lambda^\nu{}_\kappa = \eta_{\rho\kappa} \tag{1.2}$$

for the transformation matrix Λ. Infinitesimally we find from this $\Lambda^\mu{}_\nu = \delta^\mu_\nu + \omega^\mu{}_\nu + \mathcal{O}(\omega^2)$ with an antisymmetric $\omega_{\mu\nu} = -\omega_{\nu\mu}$.

In quantum theory symmetry generators are represented by unitary operators $\mathscr{U}(\Lambda)$. These furnish a representation of the Lorentz group and hence obey the composition property

$$\mathscr{U}(\Lambda)\mathscr{U}(\Lambda') = \mathscr{U}(\Lambda\Lambda'). \tag{1.3}$$

Infinitesimally we write $\mathscr{U}(\mathbb{1} + \omega) = \mathbb{1} + \frac{i}{2}\omega_{\mu\nu} M^{\mu\nu}$ with the Hermitian operators $M^{\mu\nu} = -M^{\nu\mu}$ acting on the Hilbert space of the quantum theory in question. They are the generators of the Lorentz group. We would now like to derive

J.M. Henn, J.C. Plefka, *Scattering Amplitudes in Gauge Theories*,
Lecture Notes in Physics 883, DOI 10.1007/978-3-642-54022-6_1,
© Springer-Verlag Berlin Heidelberg 2014

the Lorentz algebra, i.e. the commutation relations of the $M^{\mu\nu}$. For this consider $\mathscr{U}(\Lambda)^{-1}\mathscr{U}(\Lambda')\mathscr{U}(\Lambda) = \mathscr{U}(\Lambda^{-1}\Lambda'\Lambda)$ for the case of infinitesimal $\Lambda' = \mathbb{1} + \omega'$. Expanding to linear order in ω' on both sides of the equation for arbitrary anti-symmetric $\omega'_{\mu\nu}$ yields the transformation property of the Lorentz generator

$$\mathscr{U}(\Lambda)^{-1}M^{\mu\nu}\mathscr{U}(\Lambda) = \Lambda^{\mu}{}_{\rho}\Lambda^{\nu}{}_{\kappa}M^{\rho\kappa}. \tag{1.4}$$

Every component of $M^{\mu\nu}$ transforms with its own $\Lambda^{\mu}{}_{\nu}$ matrix. Hence we expect that a general contravariant vector P^{μ} transforms as

$$\mathscr{U}(\Lambda)^{-1}P^{\mu}\mathscr{U}(\Lambda) = \Lambda^{\mu}{}_{\nu}P^{\nu}, \tag{1.5}$$

which holds in particular for the generator of translations, the momentum operator P^{μ}, to be considered here. Taking now also the Lorentz transformation in Eqs. (1.4) and (1.5) to be infinitesimal, $\Lambda = \mathbb{1} + \omega$, and stripping off the arbitrary anti-symmetric parameter $\omega_{\rho\kappa}$ on both sides of the resulting linearized equations yields the inhomogeneous Lorentz algebra or Poincaré algebra. We find

$$\left[M^{\mu\nu}, M^{\rho\kappa}\right] = i\left(\eta^{\nu\rho}M^{\mu\kappa} + \eta^{\mu\kappa}M^{\nu\rho} - \eta^{\nu\kappa}M^{\mu\rho} - \eta^{\mu\rho}M^{\nu\kappa}\right), \tag{1.6}$$

$$\left[M^{\mu\nu}, P^{\rho}\right] = -i\eta^{\mu\rho}P^{\nu} + i\eta^{\nu\rho}P^{\mu}. \tag{1.7}$$

A general representation for the Lorentz generators is given by

$$\left(M^{\mu\nu}\right)^{i}{}_{j} = i\left(x^{\mu}\frac{\partial}{\partial x_{\nu}} - x^{\nu}\frac{\partial}{\partial x_{\mu}}\right)\delta^{i}{}_{j} + \left(S^{\mu\nu}\right)^{i}{}_{j}, \tag{1.8}$$

with x^{μ}-independent $d_R \times d_R$ representation matrices $(S^{\mu\nu})^{i}{}_{j}$ obeying the commutation relations of Eq. (1.6).

We now wish to classify the possible representations of the Lorentz group. For this define the rotation and boost generators

$$J_i := \frac{1}{2}\varepsilon_{ijk}M_{jk}, \qquad K := M_{0i}, \tag{1.9}$$

with $i, j, k = 1, 2, 3$ running over only the spatial indices. The J_i obey the $su(2)$ Lie algebra relations known from the angular momentum or spin commutation relations in quantum mechanics. Introducing the complex combinations of Hermitian generators

$$N_i := \frac{1}{2}(J_i + iK_i), \qquad N_i^{\dagger} := \frac{1}{2}(J_i - iK_i), \tag{1.10}$$

we see that the $so(1, 3)$ algebra of Eq. (1.6) may be mapped to two commuting copies of $su(2)$,

$$[N_i, N_j] = i\varepsilon_{ijk}N_k, \qquad [N_i^{\dagger}, N_j^{\dagger}] = i\varepsilon_{ijk}N_k^{\dagger}, \qquad [N_i, N_j^{\dagger}] = 0. \tag{1.11}$$

Table 1.1 Lower spin representations of the four dimensional Lorentz group. For the considerations in this text only the first four will be of importance. We have $\alpha = 1, 2$, $\dot{\alpha} = 1, 2$ and $\mu = 0, 1, 2, 3$

Rep.	Spin	Field	Lorentz transformation property
$(0,0)$	0	scalar $\phi(x)$	$\mathscr{U}(\Lambda^{-1})\phi(x)\mathscr{U}(\Lambda) = \phi(\Lambda^{-1}x)$
$(\frac{1}{2},0)$	$1/2$	left-handed Weyl spinor $\chi_\alpha(x)$	$\mathscr{U}(\Lambda^{-1})\chi_\alpha(x)\mathscr{U}(\Lambda) = L(\Lambda)_\alpha{}^\beta \chi_\beta(\Lambda^{-1}x)$
$(0,\frac{1}{2})$	$1/2$	right-handed Weyl spinor $\bar{\xi}_{\dot{\alpha}}(x)$	$\mathscr{U}(\Lambda^{-1})\bar{\xi}_{\dot{\alpha}}(x)\mathscr{U}(\Lambda) = R(\Lambda)_{\dot{\alpha}}{}^{\dot{\beta}}\bar{\xi}_{\dot{\beta}}(\Lambda^{-1}x)$
$(\frac{1}{2},\frac{1}{2})$	1	vector $A_\mu(x)$	$\mathscr{U}(\Lambda^{-1})A_\mu(x)\mathscr{U}(\Lambda) = \Lambda_\mu{}^\nu A_\nu(\Lambda^{-1}x)$
$(1,0)$	1	self-dual rank 2 tensor $B_{\mu\nu}(x)$	
$(0,1)$	1	anti-self-dual rank 2 tensor $\tilde{B}_{\mu\nu}(x)$	
$(1,\frac{1}{2})$	$3/2$	gravitino $\psi_\alpha^\mu(x)$	
$(1,1)$	2	graviton $h_{\mu\nu}(x)$	

Based on our knowledge of the representation theory of $SU(2)$ from the study of angular momentum in quantum mechanics we conclude that the representations of the $SO(1,3)$ Lorentz group may be labeled by a doublet of half-integers (m,n) related to the eigenvalues $n(n+1)$ and $m(m+1)$ of the Casimir operators $N_i N_i$ and $N_i^\dagger N_i^\dagger$ respectively. Moreover, since $J_3 = N_3 + N_3^\dagger$ we identify $m+n$ as the spin of the representation (m,n), see Table 1.1.

1.2 Weyl and Dirac Spinors

We now wish to construct a Lagrangian for the $(\frac{1}{2},0)$ field $\chi_\alpha(x)$ being the left-handed Weyl spinor of Table 1.1. The relevant 2×2 representation matrix $S_{\mathrm{L}}^{\mu\nu}$ arising in the infinitesimal Lorentz transformation $L(\mathbb{1} + \omega)_\alpha{}^\beta = \delta_\alpha^\beta + \frac{i}{2}\omega_{\mu\nu}(S_{\mathrm{L}}^{\mu\nu})_\alpha{}^\beta$ takes the form

$$\left(S_{\mathrm{L}}^{\mu\nu}\right)_\alpha{}^\beta = \frac{i}{4}\left(\sigma^\mu\bar{\sigma}^\nu - \sigma^\nu\bar{\sigma}^\mu\right)_\alpha{}^\beta, \qquad (1.12)$$

with $(\bar{\sigma}^\mu)^{\dot{\alpha}\alpha} = (\mathbb{1}, -\sigma)$ and $(\sigma^\mu)_{\alpha\dot{\alpha}} = \varepsilon_{\alpha\beta}\varepsilon_{\dot{\alpha}\dot{\beta}}(\bar{\sigma}^\mu)^{\dot{\beta}\beta} = (\mathbb{1}, \sigma)$ where σ are the Pauli matrices and $\varepsilon_{\alpha\beta}$ is the Levi-Civita tensor.[1] The free Lagrangian for a massive Weyl spinor invariant under Lorentz transformations is then

$$\mathscr{L}_{\mathrm{W}} = i\tilde{\chi}_{\dot{\alpha}}\left(\bar{\sigma}^\mu\right)^{\dot{\alpha}\alpha}\partial_\mu\chi_\alpha - \frac{1}{2}m\chi^\alpha\chi_\alpha - \frac{1}{2}m^*\tilde{\chi}_{\dot{\alpha}}\tilde{\chi}^{\dot{\alpha}}. \qquad (1.13)$$

We note $(\chi_\alpha)^\dagger = \tilde{\chi}_{\dot{\alpha}}$, $\partial_\mu = \frac{\partial}{\partial x^\mu}$ and recall that the fermionic fields are anti-commuting (Grassmann odd) quantities. The equations of motion follow from the

---

[1] Our conventions are summarized in Appendix B.

action principle by variation of the action $S = \int d^4x \mathscr{L}_W$

$$-\frac{\delta S}{\delta \tilde{\chi}_{\dot{\alpha}}} = -i(\bar{\sigma}^\mu)^{\dot{\alpha}\alpha} \partial_\mu \chi_\alpha + m^* \tilde{\chi}^{\dot{\alpha}} = 0, \tag{1.14}$$

$$-\frac{\delta S}{\delta \chi^\alpha} = -i(\sigma^\mu)_{\alpha\dot{\alpha}} \partial_\mu \tilde{\chi}^{\dot{\alpha}} + m \chi_\alpha = 0. \tag{1.15}$$

Note that the lowering of the spinor indices in the second equation introduces a minus sign in the kinetic term. In general the mass may be taken complex $m = |m|e^{i\alpha}$. However, the phase of a complex mass may be absorbed in a redefinition of the spinor fields so that we take $m = m^*$. These equations of motion may be unified into a four-component notation via

$$0 = \begin{pmatrix} m\delta_\alpha^\beta & -i(\sigma^\mu)_{\alpha\dot{\beta}} \partial_\mu \\ -i(\bar{\sigma}^\mu)^{\dot{\alpha}\beta} \partial_\mu & m\delta_{\dot{\beta}}^{\dot{\alpha}} \end{pmatrix} \begin{pmatrix} \chi_\beta \\ \tilde{\chi}^{\dot{\beta}} \end{pmatrix}. \tag{1.16}$$

Introducing the 4×4 Dirac matrices in the chiral representation

$$\gamma^\mu := \begin{pmatrix} 0 & (\sigma^\mu)_{\alpha\dot{\beta}} \\ (\bar{\sigma}^\mu)^{\dot{\alpha}\beta} & 0 \end{pmatrix}, \tag{1.17}$$

obeying the Clifford algebra $\{\gamma^\mu, \gamma^\nu\} = 2\eta^{\mu\nu}$ calls for the introduction of a four-component Majorana field $\psi_M(x)$

$$\psi_M = \begin{pmatrix} \chi_\beta \\ \tilde{\chi}^{\dot{\beta}} \end{pmatrix}, \tag{1.18}$$

whose equation of motion may be cast in the form of a Dirac equation

$$(-i\gamma^\mu \partial_\mu + m)\psi_M = 0. \tag{1.19}$$

Generalizing this, we may combine a left-handed Weyl spinor χ_α and an independent right-handed Weyl spinor $\bar{\xi}^{\dot{\alpha}}$ into a four component Dirac spinor

$$\psi = \begin{pmatrix} \chi_\alpha \\ \bar{\xi}^{\dot{\alpha}} \end{pmatrix}. \tag{1.20}$$

We can then write down the Dirac equation $(-i\gamma^\mu \partial_\mu + m)\psi = 0$. The associated Lagrangian then reads, in an index free notation,

$$\mathscr{L}_D = i\bar{\psi}\gamma^\mu \partial_\mu \psi - m\bar{\psi}\psi, \quad \text{with } \bar{\psi} := \psi^\dagger \gamma^0. \tag{1.21}$$

Exercise 1.1 (Manipulating Spinor Indices)

The ε symbols are used to raise and lower Weyl indices according to $\bar{\xi}_{\dot{\alpha}} = \varepsilon_{\dot{\alpha}\dot{\beta}} \bar{\xi}^{\dot{\beta}}$ and $\chi^\alpha = \varepsilon^{\alpha\beta} \chi_\beta$. We have

$$\varepsilon_{12} = \varepsilon_{\dot{1}\dot{2}} = \varepsilon^{21} = \varepsilon^{\dot{2}\dot{1}} = 1, \qquad \varepsilon_{21} = \varepsilon_{\dot{2}\dot{1}} = \varepsilon^{12} = \varepsilon^{\dot{1}\dot{2}} = -1.$$

The sigma matrix is defined by $(\bar{\sigma}^\mu)^{\dot{\alpha}\alpha} = (\mathbb{1}, -\sigma)$. Moreover we have $\sigma^\mu_{\alpha\dot{\alpha}} :=$ $\varepsilon_{\alpha\beta}\varepsilon_{\dot{\alpha}\dot{\beta}}\bar{\sigma}^{\mu\dot{\beta}\beta}$. Prove the relations

$$\sigma^\mu_{\alpha\dot{\alpha}} = (\mathbb{1}, \sigma), \qquad \sigma_{\mu\alpha\dot{\alpha}} = (\mathbb{1}, -\sigma),$$

$$\sigma^\mu_{\alpha\dot{\alpha}}\sigma_{\mu\beta\dot{\beta}} = 2\varepsilon_{\alpha\beta}\varepsilon_{\dot{\alpha}\dot{\beta}}, \qquad \varepsilon^{\alpha\beta}\varepsilon^{\dot{\alpha}\dot{\beta}}\sigma^\mu_{\alpha\dot{\alpha}}\sigma^\nu_{\beta\dot{\beta}} = 2\eta^{\mu\nu}.$$

Exercise 1.2 (Massless Dirac Equation and Weyl Spinors)
  Consider the (standard) representation of the Dirac matrices

$$\gamma^0 = \begin{pmatrix} \mathbb{1}_{2\times2} & 0 \\ 0 & -\mathbb{1}_{2\times2} \end{pmatrix}, \qquad \gamma^i = \begin{pmatrix} 0 & \sigma^i \\ -\sigma^i & 0 \end{pmatrix},$$

$$\gamma^5 = i\gamma^0\gamma^1\gamma^2\gamma^3 = \begin{pmatrix} 0 & \mathbb{1}_{2\times2} \\ \mathbb{1}_{2\times2} & 0 \end{pmatrix}.$$

(a) Show that the solutions of the massless Dirac equation $\gamma^\mu k_\mu \psi = 0$ may be chosen as

$$u_+(k) = v_-(k) = \frac{1}{\sqrt{2}} \begin{pmatrix} \sqrt{k^+} \\ \sqrt{k^-}e^{i\phi(k)} \\ \sqrt{k^+} \\ \sqrt{k^-}e^{i\phi(k)} \end{pmatrix},$$

$$u_-(k) = v_+(k) = \frac{1}{\sqrt{2}} \begin{pmatrix} \sqrt{k^-}e^{-i\phi(k)} \\ -\sqrt{k^+} \\ -\sqrt{k^-}e^{-i\phi(k)} \\ \sqrt{k^+} \end{pmatrix}$$

where

$$e^{\pm i\phi(k)} := \frac{k^1 \pm ik^2}{\sqrt{k^+k^-}}, \qquad k^\pm := k^0 \pm k^3,$$

and show that the spinors $u_\pm(k)$ and $v_\pm(k)$ obey the helicity relations

$$P_\pm := \frac{1}{2}(\mathbb{1} \pm \gamma^5), \qquad P_\pm u_\pm = u_\pm, \qquad P_\pm u_\mp = 0,$$

$$P_\pm v_\pm = 0, \qquad P_\pm v_\mp = v_\mp.$$

(b) What helicity relations hold for the conjugate expressions $\bar{u}_\pm(k)$ and $\bar{v}_\pm(k)$, where of course $\bar{\psi} := \psi^\dagger\gamma^0$?
(c) Now consider the unitary transformation

$$\psi \to U\psi, \qquad \gamma^\mu \to U\gamma^\mu U^\dagger,$$

using $U = \frac{1}{\sqrt{2}}(1 - i\gamma^1\gamma^2\gamma^3)$ to the chiral representation of the Dirac matrices:

$$\gamma_{ch}^0 = \begin{pmatrix} 0 & \mathbb{1}_{2\times2} \\ \mathbb{1}_{2\times2} & 0 \end{pmatrix}, \qquad \gamma_{ch}^i = \begin{pmatrix} 0 & \sigma^i \\ -\sigma^i & 0 \end{pmatrix}, \qquad i = 1, 2, 3.$$

Determine γ^5 and the spinors $u_\pm(k)$ and $v_\pm(k)$ in this chiral basis!

1.3 Non-Abelian Gauge Theories

We now discuss the enormously important principle of local gauge invariance due to Yang and Mills [1] leading to non-Abelian gauge theories. The fermion Lagrangians $\mathscr{L}_W$ of Eq. (1.13) and $\mathscr{L}_D$ of Eq. (1.21) are invariant under global phase transformations

$$\chi_\alpha \to e^{i\alpha}\chi_\alpha, \quad \text{respectively} \quad \psi \to e^{i\alpha}\psi, \quad \alpha \in \mathbb{R}. \tag{1.22}$$

However, for the case of a *local* phase transformation $\alpha(x)$ with arbitrary space-time dependence, the kinetic terms in the actions $\mathscr{L}_W$ and $\mathscr{L}_D$ are no longer invariant. This may be cured by introducing a gauge field $A_\mu(x)$ coupling to the spinor fields in a unique fashion to cancel these unwanted terms. For this the local gauge transformations of the Dirac field (we specialize to this case from now on) and the novel gauge field $A_\mu(x)$ are introduced

$$\psi \to e^{ie\alpha(x)}\psi, \qquad A_\mu \to A_\mu + \partial_\mu\alpha(x), \tag{1.23}$$

and the derivative ∂_μ in the Dirac action is replaced by the covariant derivative $D_\mu = \partial_\mu - ieA_\mu$, with e denoting the coupling constant. The Dirac Lagrangian is then lifted to the locally gauge invariant Lagrangian

$$\mathscr{L}_{QED} = i\bar{\psi}\gamma^\mu D_\mu\psi - m\bar{\psi}\psi - \frac{1}{4}F_{\mu\nu}F^{\mu\nu}, \tag{1.24}$$

which for the case of e and m being the charge and mass of the electron is the theory of electrodynamics. This is an Abelian gauge theory. The field strength tensor generating the kinetic term for the gauge field A_μ above is defined as

$$F_{\mu\nu} = \frac{i}{e}[D_\mu, D_\nu] = \partial_\mu A_\nu - \partial_\nu A_\mu. \tag{1.25}$$

Let us now formalize this construction slightly by associating to the local Abelian gauge transformation of Eq. (1.23) an x-dependent $U(1)$ group element $U(x) = e^{ie\alpha(x)}$ obeying $U(x)^\dagger U(x) = 1$ and generating the transformations

$$\psi \to U(x)\psi, \qquad D_\mu \to U(x)D_\mu U^\dagger(x), \tag{1.26}$$

which leave $\bar{\psi}\gamma^\mu D_\mu \psi$ and $\bar{\psi}\psi$ manifestly invariant, as $\bar{\psi} \to \bar{\psi}U^{-1}(x)$. The transformation rule for $D_\mu = \partial_\mu - ieA_\mu$ above implies the transformation of the gauge field

$$A_\mu(x) \to U(x)A_\mu(x)U^\dagger(x) + \frac{i}{e}U(x)\partial_\mu U^\dagger(x). \tag{1.27}$$

One indeed easily verifies the equivalence to Eq. (1.23).

We now wish to lift this construction to a non-Abelian gauge symmetry. For this we consider a set of N Dirac spinor fields $\psi_{i,A}$ with spinor index $A = (\alpha, \dot{\alpha})$ and $i = 1, \ldots, N$. The associated Dirac-Lagrangian

$$\mathscr{L}_{D_N} = \sum_{i=1}^N i\bar{\psi}^i \slashed{\partial}\psi_i - m\bar{\psi}^i\psi_i \tag{1.28}$$

is now invariant under the global unitary transformation

$$\psi_i(x) \to U_i^j\psi_j(x) \tag{1.29}$$

with the $N \times N$ matrices U_j^i obeying $U^\dagger U = UU^\dagger = \mathbb{1}$ spanning the $U(N)$ Lie group. We shall further specialize to the case of special unitary transformations $SU(N)$ with the additional condition $\det(U) = 1$.

In these lectures we will focus on gauge theories built from $SU(N)$, although all other compact semi-simple Lie groups $SO(N)$, $Sp(2N)$ and the five exceptional G_2, F_4, E_6, E_7 and E_8 may in principle be also considered.[2]

The global symmetry of $\mathscr{L}_{D_N}$ under Eq. (1.29) may now be turned into a *local* non-Abelian symmetry $U_i^j \to U_i^j(x)$ with arbitrary space-time dependence by following the construction above. We introduce the covariant $N \times N$ matrix valued derivative,

$$(D_\mu)_i^j = \delta_i^j \partial_\mu - ig(A_\mu)_i^j(x), \tag{1.30}$$

with the $SU(N)$ gauge field $(A_\mu)_i^j(x)$, which is a Hermitian and traceless matrix. The coupling constant is denoted by g. Demanding a covariant transformation generalizing Eq. (1.26)

$$(D_\mu)_i^j \to U_i^k(x)(D_\mu)_k^l(U^\dagger)_l^j(x) \tag{1.31}$$

leads to (now in matrix notation)

$$A_\mu(x) \to U(x)A_\mu(x)U^\dagger(x) + \frac{i}{g}U(x)\partial_\mu U^\dagger(x). \tag{1.32}$$

With the help of such a construction the 'gauged' Lagrangian

$$\mathscr{L}'_{D_n} = i\bar{\psi}^i \slashed{D}_i^j \psi_j - m\bar{\psi}^i\psi_i \tag{1.33}$$

---

[2]However, differences arise in the large N limit to be discussed below, which of course does not exist for the exceptional groups.

is manifestly invariant under *local SU(N)* gauge transformations. What is still missing is a kinetic term for the non-Abelian gauge field $(A_\mu)_i^j$. In generalization of the Abelian construction of Eq. (1.25) the natural quantity to take is in matrix notation

$$F_{\mu\nu} = \frac{i}{g}[D_\mu, D_\nu] = \partial_\mu A_\nu - \partial_\nu A_\mu - ig[A_\mu, A_\nu], \tag{1.34}$$

which opposed to the Abelian case now transforms covariantly under gauge transformations, i.e. we have $F_{\mu\nu} \to U(x)F_{\mu\nu}U^\dagger(x)$. It can be made gauge invariant by taking a trace. Hence the kinetic term for the gauge field is

$$\mathcal{L}_{YM} = -\frac{1}{4}\operatorname{Tr}\left(F_{\mu\nu}F^{\mu\nu}\right) \tag{1.35}$$

being both gauge and Lorentz invariant. We note that opposed to the $U(1)$ case the non-Abelian gauge field is self-interacting due to the commutator term in Eq. (1.34). The interaction strength is controlled by the coupling constant g. The complete Lagrangian of $SU(N)$ gauge theory interacting with a Dirac 'matter' field is then given by the sum of $\mathcal{L}_{D_n}'$ and $\mathcal{L}_{YM}$.

In order to better understand the structure of this gauge theory it is useful to look at gauge transformations infinitesimally close to the identity

$$U_{ij}(x) = \delta_i^j - ig\theta^a(x)(T^a)_i^j, \tag{1.36}$$

introducing the Lie algebra generators $(T^a)_j^i$ of $SU(N)$ with $a = 1, \ldots, N^2 - 1$ and $i, j = 1, \ldots, N$. In the above the $\theta^a(x)$ serve as the local transformation parameters generalizing the phase $\alpha(x)$ for the Abelian case $N = 1$. In consequence of the $SU(N)$ group properties the $(T^a)_i^j$ are Hermitian traceless $N \times N$ matrices. They obey the commutation relations

$$\left[T^a, T^b\right] = i\sqrt{2}f^{abc}T^c, \tag{1.37}$$

with the gauge group structure constants f^{abc}. The factor $\sqrt{2}$ is our normalization convention. We can choose a diagonal basis for the generators T^a such that

$$\operatorname{Tr}\left(T^aT^b\right) = \delta^{ab}. \tag{1.38}$$

Based on this we may write

$$f^{abc} = -\frac{i}{\sqrt{2}}\operatorname{Tr}\left(T^a\left[T^b, T^c\right]\right), \tag{1.39}$$

which renders the structure constants totally anti-symmetric in all indices. Furthermore we note the important $SU(N)$ identity

$$(T^a)_{i_1}^{j_1}(T^a)_{i_2}^{j_2} = \delta_{i_1}^{j_2}\delta_{i_2}^{j_1} - \frac{1}{N}\delta_{i_1}^{j_1}\delta_{i_2}^{j_2}, \tag{1.40}$$

which is nothing but a completeness relation for $N \times N$ Hermitian matrices and where the last term ensures the tracelessness of the $(T^a)_i^j$. Concrete expressions for the $SU(N)$ generators are:

- $SU(2)$: $T^a = \frac{1}{\sqrt{2}}\sigma^a$, σ^a: Pauli matrices $a = 1, 2, 3$.
- $SU(3)$: $T^a = \frac{1}{\sqrt{2}}\lambda^a$, λ^a: Gell-Mann matrices $a = 1, \ldots, 8$.

$$\lambda^a = \begin{pmatrix} \sigma^a & \\ & 0 \end{pmatrix}, \qquad \lambda^4 = \begin{pmatrix} 0 & & 1 \\ & 0 & \\ 1 & & 0 \end{pmatrix}, \qquad \lambda^5 = \begin{pmatrix} 0 & & -i \\ & 0 & \\ i & & 0 \end{pmatrix},$$

$$\lambda^6 = \begin{pmatrix} 0 & & \\ & 0 & 1 \\ & 1 & 0 \end{pmatrix}, \qquad \lambda^7 = \begin{pmatrix} 0 & & \\ & 0 & -i \\ & i & 0 \end{pmatrix}, \qquad (1.41)$$

$$\lambda^8 = \frac{1}{\sqrt{3}} \begin{pmatrix} 1 & & \\ & 1 & \\ & & -2 \end{pmatrix}.$$

- $SU(N)$: Explicit constructions exist e.g. in terms of the 't Hooft twist matrices [2].

As we took the gauge fields to be traceless Hermitian matrices we can expand them in the basis of $SU(N)$ generators T^a

$$(A_\mu)_i^j(x) = A_\mu^a(x)(T^a)_i^j \quad \Leftrightarrow \quad A_\mu^a(x) = \mathrm{Tr}(T^a A_\mu(x)). \qquad (1.42)$$

Similarly the field strength may be decomposed as $F_{\mu\nu}(x) = F_{\mu\nu}^a(x)T^a$ and

$$F_{\mu\nu}^c = \partial_\mu A_\nu^c - \partial_\nu A_\mu^c + \sqrt{2}g f^{abc} A_\mu^a A_\nu^b, \qquad (1.43)$$

yielding $\mathscr{L}_{YM} = -\frac{1}{4}F_{\mu\nu}^c F^{c\mu\nu}$.

An important example is Quantum Chromodynamics (QCD), the theory of the strong interactions, which has the gauge group $SU(3)$. The field content consists of 8 gauge fields known as gluons, $A_\mu^a(x)$ ($a = 1, \ldots, 8$), together with 6 flavors of quark fields, $\psi_{i,I}$ ($i = 1, 2, 3$; $I = 1, \ldots, 6$) known as up, down, strange, charm, bottom and top quarks. Its Lagrangian reads

$$\mathscr{L}_{QCD} = i\bar{\psi}_I^i \not{D}_i^j \psi_{Ij} - m_I \bar{\psi}_I^i \psi_{Ii} - \frac{1}{4}F_{\mu\nu}^c F^{c\mu\nu}, \qquad (1.44)$$

where the masses m_I span a range from 2 MeV for the up-quark to 172 GeV for the top-quark.

Given the structure constants f^{abc} of a non-Abelian gauge group as in Eq. (1.37) one can search for representations of the group in terms of $d_R \times d_R$ dimensional matrices $(T_R^a)_I^J$ with $I, J = 1, \ldots, d_R$ obeying

$$[T_R^a, T_R^b] = i\sqrt{2} f^{abc} T_R^c, \qquad (1.45)$$

with d_R denoting the dimension of the representation. So far we discussed the fundamental or defining representation of $SU(N)$ in the form of $N \times N$ Hermitian matrices with the explicit forms spelled out in Eq. (1.41). As f^{abc} is real we see by complex conjugating Eq. (1.45) that for a given representation T_R^a there always exists a complex conjugate representation $T_{\bar{R}}^a := -T_R^{a*}$. An important representation next to the fundamental one is the $N^2 - 1$ dimensional adjoint representation induced by the structure constants f^{abc} of the algebra as follows,

$$\left(T_A^a\right)^{bc} := -i\sqrt{2} f^{abc}, \tag{1.46}$$

where Eq. (1.45) is a consequence of the Jacobi identity for f^{abc}. The matrix indices are raised and lowered freely in this representation due to the diagonal metric Eq. (1.38). We also note that $T_{\bar{A}}^a = T_A^a$, as f^{abc} is real. Indeed, the infinitesimal transformation of gauge fields following from Eqs. (1.32) and (1.36) reads

$$\delta A_\mu^a = \partial_\mu \Theta^a + \sqrt{2} g f^{abc} \Theta^b A_\mu^c = \partial_\mu \Theta^a + ig\Theta^b \left(T_A^b\right)^{ac} A_\mu^c. \tag{1.47}$$

Quarks on the other hand transform in the N dimensional fundamental representation

$$\delta\psi_i = \Theta^a \left(T_N^a\right)_i^j \psi_j. \tag{1.48}$$

Comparing the two, we see that gauge fields transform in the adjoint representation in their homogeneous part. As a final comment representations are often denoted by their dimensionality in boldfaced letters, e.g. for QCD the quarks are in the **3**, the anti-quarks in the **3̄** whereas the gluons are in the **8** of $SU(3)$.

1.4 Scattering Amplitudes and Feynman Rules for Gauge Theories

The central experimental observables in particle physics are cross sections for scattering processes as they occur in colliders. These are derived from the associated scattering amplitudes which correspond to the probability amplitude for the specific process to occur. The amplitude for the scattering of an initial state at $t \to -\infty$ of n_i-particles into a final state at $t \to +\infty$ of n_f particles is determined by the S-matrix element $\langle \text{out}|S|\text{in}\rangle$. The asymptotic states are well separated and non-interacting and given by the direct product of single particle states of definite on-shell momentum and polarization denoted by $|\{p_i, \varepsilon_i\}\rangle$ with $p_i^2 = m_i^2$. The scattering amplitude reads

$$\mathscr{A}_{n_i \to n_f} = \langle \{q_1, \varepsilon_1'\}, \dots, \{q_{n_f}, \varepsilon_{n_f}'\} | S | \{p_1, \varepsilon_1\}, \dots, \{p_{n_i}, \varepsilon_{n_i}\}\rangle. \tag{1.49}$$

The observable differential cross section is proportional to $|\mathscr{A}_{n_i \to n_f}|^2$, the proportionality factor being a kinematical function of the involved $n_i + n_f$ particles. It is derived in standard text books on quantum field theory. A central result in quantum field theory is the Lehman-Symanzik-Zimmermann reduction formula, which relates the scattering amplitude $\mathscr{A}_{n_i \to n_f}$ to the renormalized Green function of $n_i + n_f$ local fields of given momenta (specializing to scalar quantum field theory of mass m for now, where there are no polarizations ε_i),

$$\mathscr{A}_{n_i \to n_f} = \prod_{i=1}^{n_f} \frac{q_i^2 - m^2 + i0}{i\sqrt{Z}} \prod_{j=1}^{n_i} \frac{p_j^2 - m^2 + i0}{i\sqrt{Z}}$$

$$\times \left. G(-q_1, \ldots, -q_m, p_1, \ldots, p_n) \right|_{q_i^2 = p_j^2 = m^2} . \tag{1.50}$$

The momentum space Green functions above may be defined via the functional (or path) integral as

$$G(k_1, k_2, \ldots, k_n) := \left\langle \Phi(k_1) \Phi(k_2) \cdots \Phi(k_n) \right\rangle$$

$$= \frac{\int [d\Phi] \Phi(k_1) \Phi(k_2) \cdots \Phi(k_n) e^{iS[\Phi]}}{\int [d\Phi] e^{iS[\Phi]}}, \tag{1.51}$$

where $\Phi(k)$ denotes a generic scalar quantum field of momentum k. It is obtained from $\Phi(x)$ via a Fourier transformation. Moreover the $\sqrt{Z}$ factors appearing in Eq. (1.50) are the *field or wave function renormalization constants* arising as the residue of the on-shell pole of the two-point Green function

$$G(p, -p) \underset{p^2 \to m^2}{\sim} \frac{iZ}{p^2 - m^2 + i0}, \tag{1.52}$$

where m_i is the physical (renormalized) mass of the particle and $i0$ represents Feynman's $i\varepsilon$ prescription to define the propagator. In an interacting theory the Green functions can be evaluated in perturbation theory upon expanding out the interaction terms in the action $S = S_0 + g \int d^4x \, \mathscr{L}_{\text{int}}$ governed by a coupling constant g which give rise to the vertex functions of the Feynman rules. Then the evaluation of $G(k_1, k_2, \ldots, k_n)$ is rewritten as a power series in the coupling constant g with each term given by a suitable product of Green functions of the *free* theory integrated over the vertex positions. The free Green functions in momentum space are nothing but products of propagators $\frac{1}{p^2 - m^2 + i0}$ and one is left to integrate over intermediate loop momenta which arise from the vertex expansion. Note that the prefactors in Eq. (1.50) precisely cancel the external leg propagators, yielding the representation of the amplitude as an amputated Green function. In a scattering process there is always the probability of having no interactions at all. One therefore splits the S-matrix into a trivial and an interacting piece

$$S = \mathbb{1} + iT. \tag{1.53}$$

It can then be shown that matrix elements of T follow from the amputated and connected Green function

$$\langle q_1, \ldots, q_{n_f} | T | p_1, \ldots, p_{n_i} \rangle$$

$$= \left(\prod_{i=1}^{n_i+m_f} \sqrt{Z_i} \right) \begin{pmatrix} \text{Sum of all amputated and connected graphs with} \\ \{p_i\} \text{ incoming and } \{q_i\} \text{ outgoing legs} \end{pmatrix}. \quad (1.54)$$

This structure remains essentially the same also for the case of fermions or gauge bosons in the asymptotic states. For these spin-1/2 and spin-1 states one then needs to specify in addition the polarization degrees of freedom on each leg: In the case of (massless) gauge fields these are the ± 1 helicity states $\varepsilon_\pm^\mu$, in the case of Dirac fields these are the fermion and anti-fermion solutions to the Dirac equation $u_s(p)$ and $v_s(p)$ with spin (or helicity in the massless case) $s = \pm 1/2$. E.g. for a helicity conserving $2 \to 2$ gluon amplitude we would have

$$G_{g+g+\to g+g+} = \varepsilon_+^{*\mu_1} \varepsilon_+^{*\mu_2} \varepsilon_+^{\nu_1} \varepsilon_+^{\nu_2} \langle A_{\mu_1}(-q_1) A_{\mu_2}(-q_2) A_{\nu_1}(p_1) A_{\nu_2}(p_2) \rangle. \quad (1.55)$$

An important further point in the case of gauge fields is the need to fix a gauge in order to render the path integral well defined. Briefly stated the divergence arises due to integrating over an infinite number of gauge equivalent field configurations. Gauge fixing selects one field configuration for each gauge orbit. This is elegantly achieved by the Fadeev-Popov procedure by adding a gauge fixing and a ghost term to the action.

After this brief review we are in a position to write down the Feynman rules of non-Abelian gauge theories. In the pure Yang-Mills sector the Lagrangian reads

$$\mathscr{L}_{YM} = -\frac{1}{4} F_{\mu\nu}^a F^{a\mu\nu}$$

$$= -\frac{1}{2} \partial^\mu A_\nu^a \partial_\mu A^{a\nu} + \frac{1}{2} \partial^\mu A_\nu^a \partial_\nu A^{a\mu}$$

$$- g f^{abc} A^{a\mu} A^{b\nu} \partial_\mu A_\nu^c - \frac{1}{4} g^2 f^{abe} f^{cde} A^{a\mu} A^{b\nu} A_\mu^c A_\nu^d. \quad (1.56)$$

The Lagrangian is not yet complete, as we need to add a gauge fixing term and ghosts. Adding the common ξ-dependent gauge fixing term

$$\mathscr{L}_{g.f.} = -\frac{1}{2\xi} \left(\partial^\mu A_\mu^a \right)^2 \quad (1.57)$$

the kinetic term of the Lagrangian becomes

$$\mathscr{L}_{YM} + \mathscr{L}_{g.f.}|_{A^2} = \frac{1}{2} A^{a\mu} \left(\eta_{\mu\nu} \partial^2 - \left(1 - \frac{1}{\xi} \right) \partial_\mu \partial_\nu \right) A^{a\nu}. \quad (1.58)$$

Table 1.2 Momentum space Feynman rules for gluons and massive Dirac fermions in the representation R. We will specialize to the Feynman gauge $\xi = 1$

Propagators

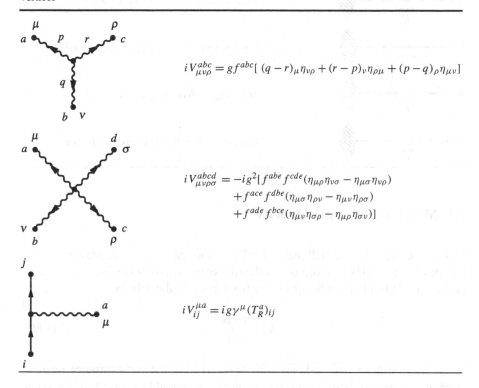

$$\Delta_{\mu\nu}^{ab}(k) = \frac{\delta^{ab}}{k^2+i0}\left(\eta_{\mu\nu} + (\xi - 1)\frac{k_\mu k_\nu}{k^2}\right)$$

$$S_{ij}(k) = \frac{\not{k}-m}{k^2-m^2+i0}\delta_{ij}$$

Vertices

$$iV_{\mu\nu\rho}^{abc} = gf^{abc}[\,(q-r)_\mu\eta_{\nu\rho} + (r-p)_\nu\eta_{\rho\mu} + (p-q)_\rho\eta_{\mu\nu}]$$

$$iV_{\mu\nu\rho\sigma}^{abcd} = -ig^2[f^{abe}f^{cde}(\eta_{\mu\rho}\eta_{\nu\sigma} - \eta_{\mu\sigma}\eta_{\nu\rho})$$
$$+ f^{ace}f^{dbe}(\eta_{\mu\sigma}\eta_{\rho\nu} - \eta_{\mu\nu}\eta_{\rho\sigma})$$
$$+ f^{ade}f^{bce}(\eta_{\mu\nu}\eta_{\sigma\rho} - \eta_{\mu\rho}\eta_{\sigma\nu})]$$

$$iV_{ij}^{\mu a} = ig\gamma^\mu(T_R^a)_{ij}$$

The ghost Lagrangian then follows from the Fadeev-Popov procedure and takes the form

$$\mathscr{L}_{\text{ghosts}} = -\partial_\mu b^a D^\mu c^a. \tag{1.59}$$

It will, however, play *no* role for us, as we shall be discussing the construction of loop amplitudes through unitarity methods. Here the only input needed are tree-level amplitudes with physical external states, which do not include ghosts. The momentum space Feynman-rules following from Eqs. (1.44), (1.56) and (1.58) are detailed in Tables 1.2 and 1.3.

Table 1.3 Polarization rules

Polarizations

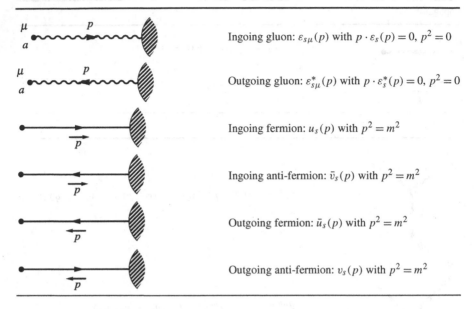

	Ingoing gluon: $\varepsilon_{s\mu}(p)$ with $p \cdot \varepsilon_s(p) = 0$, $p^2 = 0$
	Outgoing gluon: $\varepsilon_{s\mu}^*(p)$ with $p \cdot \varepsilon_s^*(p) = 0$, $p^2 = 0$
	Ingoing fermion: $u_s(p)$ with $p^2 = m^2$
	Ingoing anti-fermion: $\bar{v}_s(p)$ with $p^2 = m^2$
	Outgoing fermion: $\bar{u}_s(p)$ with $p^2 = m^2$
	Outgoing anti-fermion: $v_s(p)$ with $p^2 = m^2$

1.5 Massless Particles: Helicity

In these lecture notes we shall mainly deal with the scattering of massless particles. For massless particles the projection of the spin onto the axis of the three-momentum of the particle is a Lorentz invariant quantity known as the helicity

$$h := \frac{\mathbf{p} \cdot \mathbf{S}}{|\mathbf{p}|}. \tag{1.60}$$

The reason for this projection being Lorentz invariant is that as the particles are traveling at the speed of light there is no Lorentz boost possible to invert the direction of propagation. A spin $1/2$ fermion takes helicity values $h = \pm 1/2$. In the massless limit the initially independent positive $u(p)$ and negative $v(p)$ energy solutions of the Dirac equation

$$(\not{p} - m)u(p) = 0, \qquad (\not{p} + m)v(p) = 0, \tag{1.61}$$

degenerate and fall together. Then the solution of definite helicity $u_\pm(p) = \frac{1}{2}(1 \pm \gamma_5)u(p)$ and $v_\mp(k) = \frac{1}{2}(1 \pm \gamma_5)v(p)$ may be identified with each other. Hence every massless fermion external leg in a scattering amplitude carries the on-shell labels $\{p_i, \pm \frac{1}{2}\}$.

Similarly spin 1 gauge fields carry helicities $h = \pm 1$ with the associated polarization vectors $\varepsilon_{\pm}^{\mu}(p)$ obeying

$$p \cdot \varepsilon_{\pm}(p) = 0, \qquad \varepsilon_{+}(p) \cdot \varepsilon_{-}(p) = -1,$$

$$\varepsilon_{+}(p) \cdot \varepsilon_{+}(p) = 0 = \varepsilon_{-}(p) \cdot \varepsilon_{-}(p), \qquad \left(\varepsilon_{\pm}^{\mu}\right)^{*} = \varepsilon_{\mp}^{\mu}. \tag{1.62}$$

Thus an external gluon leg carries the on-shell labels $\{p_i, \pm 1\}$.

1.6 Spinor Helicity Formalism for Massless Particles

We shall now introduce the highly useful spinor helicity formalism for the description of scattering amplitudes of massless particles. It provides a uniform description of the on-shell degrees of freedom (momentum and polarization) for the scattering states of all helicities (gluons, fermions, scalars) of massless particles. It also renders the analytic expressions of scattering amplitudes in an often much more compact form compared to the standard four-vector notation.

To begin with we write the four-momentum p^{μ} of an on-shell state as a matrix in spinor indices via

$$p^{\mu} \rightarrow p^{\dot{\alpha}\alpha} = \bar{\sigma}_{\mu}^{\dot{\alpha}\alpha} p^{\mu} = \begin{pmatrix} p^0 + p^3 & p^1 - ip^2 \\ p^1 + ip^2 & p^0 - p^3 \end{pmatrix}, \qquad \bar{\sigma}_{\mu}^{\dot{\alpha}\alpha} = (\mathbb{1}, \sigma). \tag{1.63}$$

Then we have

$$p_i^{\dot{\alpha}\alpha} p_j^{\dot{\beta}\beta} \varepsilon_{\alpha\beta} \varepsilon_{\dot{\alpha}\dot{\beta}} = p_i^{\mu} p_j^{\nu} \underbrace{\bar{\sigma}_{\mu}^{\dot{\alpha}\alpha} \bar{\sigma}_{\nu}^{\dot{\beta}\beta} \varepsilon_{\alpha\beta} \varepsilon_{\dot{\alpha}\dot{\beta}}}_{=2\eta_{\mu\nu}} = 2p_i \cdot p_j, \tag{1.64}$$

and the mass-shell condition may be reexpressed as a determinant condition on $p_i^{\dot{\alpha}\alpha}$

$$p_i^2 = m^2 \quad \Leftrightarrow \quad p_i \cdot p_i = \frac{1}{2} p_i^{\dot{\alpha}\alpha} p_i^{\dot{\beta}\beta} \varepsilon_{\alpha\beta} \varepsilon_{\dot{\alpha}\dot{\beta}} = \det\left(p_i^{\dot{\alpha}\alpha}\right) = m^2. \tag{1.65}$$

Any 2×2 matrix has at most rank two and therefore may be written as $p^{\alpha\dot{\alpha}} = \lambda^{\alpha}\tilde{\lambda}^{\dot{\alpha}} + \mu^{\alpha}\tilde{\mu}^{\dot{\alpha}}$ with commuting Weyl spinors λ, μ and $\tilde{\lambda}, \tilde{\mu}$ in the $(0, 1/2)$ and $(1/2, 0)$ representations of the Lorentz-group, respectively. We now specialize to massless particles

$$\det\left(p_i^{\dot{\alpha}\alpha}\right) = 0. \tag{1.66}$$

The rank of a 2×2 matrix is less than two if its determinant vanishes and hence the light-like four-momenta are precisely those which may be written as

$$p_i^{\dot{\alpha}\alpha} = \tilde{\lambda}_i^{\dot{\alpha}} \lambda_i^{\alpha}, \tag{1.67}$$

representing a rank-one 2×2 matrix. The λ_i^α and $\tilde{\lambda}_i^{\dot\alpha}$ are known as the helicity spinors. Requiring the four-momentum to be real with Lorentz-signature translates into the relation $(\lambda_i^\alpha)^* = \pm \tilde{\lambda}_i^{\dot\alpha}$. The sign in this relation follows the sign of the energy of the associated four-momentum. An explicit realization is

$$\lambda^\alpha = \frac{1}{\sqrt{p^0 + p^3}} \begin{pmatrix} p^0 + p^3 \\ p^1 + ip^2 \end{pmatrix}, \qquad \tilde{\lambda}^{\dot\alpha} = \frac{1}{\sqrt{p^0 + p^3}} \begin{pmatrix} p^0 + p^3 \\ p^1 - ip^2 \end{pmatrix}. \qquad (1.68)$$

Note that $\sqrt{p^0 + p^3}$ is real (imaginary) for positive (negative) p^0 by virtue of $|p^0| \geq |p^3|$. If one extends the definition of the momenta into the complex plane (which we shall do in the following) then the spinors λ_i and $\tilde{\lambda}_i$ are independent. Spinor indices are raised and lowered with the invariant Levi-Civita tensor,

$$\lambda_\alpha := \varepsilon_{\alpha\beta} \lambda^\beta, \qquad \tilde{\lambda}_{\dot\alpha} := \varepsilon_{\dot\alpha\dot\beta} \tilde{\lambda}^{\dot\beta}. \qquad (1.69)$$

In the spinor helicity formalism a general scattering amplitude involving massless particles is a function of the set of helicity spinors $\{\lambda_i, \tilde{\lambda}_i\}$ with the index i running over all in- and outgoing legs. In order to uniformize the description we shall take all particles as outgoing from now on. Lorentz-invariant quantities may be built from the λ_i and $\tilde{\lambda}_i$ through the angle $\langle .. \rangle$ and square $[..]$ brackets as the basic spinor invariants

$$\langle \lambda_i \lambda_j \rangle := \lambda_i^\alpha \lambda_{j\alpha} = \varepsilon_{\alpha\beta} \lambda_i^\alpha \lambda_j^\beta = -\langle \lambda_j \lambda_i \rangle =: \langle ij \rangle,$$
$$[\tilde{\lambda}_i \tilde{\lambda}_j] := \tilde{\lambda}_{i\dot\alpha} \tilde{\lambda}_j^{\dot\alpha} = -\varepsilon_{\dot\alpha\dot\beta} \tilde{\lambda}_i^{\dot\alpha} \tilde{\lambda}_j^{\dot\beta} = -[\lambda_j \lambda_i] =: [ij], \qquad (1.70)$$

where $i, j = 1, \ldots, n$. Note the opposite index contraction conventions for left-handed and right-handed Weyl spinors. Also note that $\langle \lambda_i \lambda_i \rangle = \langle ii \rangle = 0$ and $[\lambda_i \lambda_i] = [ii] = 0$.

In terms of these brackets the Mandelstam invariants s_{ij} may be written as

$$s_{ij} = (p_i + p_j)^2 = 2p_i \cdot p_j = p_i^{\alpha\dot\alpha} p_{j\alpha\dot\alpha} = \langle ij \rangle [ji]. \qquad (1.71)$$

From this we see that for real momenta we have

$$\langle ij \rangle = \sqrt{|s_{ij}|} e^{i\phi_{ij}}, \qquad [ij] = -\sqrt{|s_{ij}|} e^{-i\phi_{ij}}, \qquad \phi_{ij} \in \mathbb{R}. \qquad (1.72)$$

Indeed for positive energy states $p_i^0 > 0$ and $p_j^0 > 0$ one shows that the phase ϕ_{ij} follows from the two momenta p_i and p_j as

$$\cos\phi_{ij} = \frac{p_i^1 p_j^+ - p_j^1 p_i^+}{\sqrt{|s_{ij}|} \sqrt{p_i^+ p_j^+}}, \qquad \sin\phi_{ij} = \frac{p_i^2 p_j^+ - p_j^2 p_i^+}{\sqrt{|s_{ij}|} \sqrt{p_i^+ p_j^+}},$$
$$p_i^\pm = p_i^0 \pm p_i^3. \qquad (1.73)$$

Note that for a given four-momentum vector $p_i^{\alpha\dot\alpha}$ the associated helicity spinors λ_i^α and $\tilde\lambda_i^{\dot\alpha}$ are only fixed up to a convention dependent phase φ_i, since the rescaling

$$\lambda_i^\alpha \to e^{-i\varphi_i}\lambda_i^\alpha, \qquad \tilde\lambda_i^{\dot\alpha} \to e^{i\varphi_i}\tilde\lambda_i^{\dot\alpha}, \tag{1.74}$$

leaves $p_i^{\alpha\dot\alpha}$ invariant, cf. Eq. (1.67). Indeed this rescaling freedom is generated by a $U(1)$ generator of *helicity*

$$h = \frac{1}{2}\sum_{i=1}^{n}\left[-\lambda_i^\alpha\frac{\partial}{\partial\lambda_i^\alpha} + \tilde\lambda_i^{\dot\alpha}\frac{\partial}{\partial\tilde\lambda_i^{\dot\alpha}}\right]. \tag{1.75}$$

With Eq. (1.75) we have assigned helicity $-\frac{1}{2}$ to λ and helicity $\frac{1}{2}$ to $\tilde\lambda$. Thus the external helicity spinor states can carry information on both the momentum and the helicity of the associated external leg in a scattering amplitude.

It is easy to see that the helicity spinors solve the massless Dirac equation. In the chiral representation of the Dirac-matrices of Eq. (1.17) we note

$$\not p = p_\mu\gamma^\mu = \begin{pmatrix} 0 & p_{\alpha\dot\alpha} \\ p^{\dot\alpha\alpha} & 0 \end{pmatrix}$$

using $\quad p_{\alpha\dot\alpha} := p_\mu(\sigma^\mu)_{\alpha\dot\alpha}, \; p^{\dot\alpha\alpha} = \varepsilon^{\alpha\beta}\varepsilon^{\dot\alpha\dot\beta}p_{\beta\dot\beta} = p_\mu(\bar\sigma^\mu)^{\dot\alpha\alpha}. \tag{1.76}$

The helicity spinors λ_α and $\tilde\lambda^{\dot\alpha}$ then appear as building blocks for the solutions of the massless Dirac equation

$$u_+(p) = v_-(p) = \begin{pmatrix} \lambda_\alpha \\ 0 \end{pmatrix} =: |p\rangle, \qquad u_-(p) = v_+(p) = \begin{pmatrix} 0 \\ \tilde\lambda^{\dot\alpha} \end{pmatrix} =: |p], \tag{1.77}$$

introducing a convenient square and angle bracket notation $|\ldots\rangle$ and $|\ldots]$, manifestly satisfying $\not p|p\rangle = 0 = \not p|p]$.

Finally we note two important identities for helicity spinors. The first is the Schouten identity

$$\langle\lambda_1\lambda_2\rangle\lambda_3^\alpha + \langle\lambda_2\lambda_3\rangle\lambda_1^\alpha + \langle\lambda_3\lambda_1\rangle\lambda_2^\alpha = 0, \tag{1.78}$$

or

$$\langle 12\rangle\langle 3n\rangle + \langle 23\rangle\langle 1n\rangle + \langle 31\rangle\langle 2n\rangle = 0, \tag{1.79}$$

with arbitrary spinor λ_n and similarly for the conjugated spinors. The second identity makes use of the total momentum conservation in scattering amplitudes (recall our convention of all legs outgoing, as depicted in Fig. 1.1)

$$\sum_{i=1}^{n}p_i^\mu = 0 \quad\Leftrightarrow\quad \sum_{i=1}^{n}\lambda_i^\alpha\tilde\lambda_i^{\dot\alpha} = 0 \quad\Leftrightarrow\quad \sum_{i=1}^{n}\langle ai\rangle[ib] = 0, \tag{1.80}$$

for arbitrary λ_a and $\tilde\lambda_b$.

Fig. 1.1 In our convention
all particles are outgoing

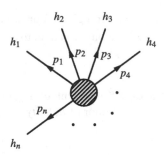

1.7 Gluon Polarizations

We now proceed to establish a bi-spinor representation for the polarization vector
of a massless gauge boson of helicity ± 1. The properties of the gluon polarization
vectors were discussed in Eq. (1.62). An external gluon leg depends on momentum,
helicity and color, which we shall compactly denote as $\{i^{\pm}, a_i\}$ with $i^{\pm} = \{p_i, h_i = \pm 1\}$ and a_i the color index. For example a pure gluon n-particle amplitude with a
specific distribution of helicities reads

$$\mathscr{A}_n\left(1^+, a_1; 2^-, a_2, 3^+, a_3; \ldots, n^-, a_n\right)$$
$$= \varepsilon_+^{\mu_1} \varepsilon_-^{\mu_2} \varepsilon_+^{\mu_3} \cdots \varepsilon_-^{\mu_n} \mathscr{A}_{\mu_1 \mu_2 \cdots \mu_n}^{a_1 a_2 \cdots a_n}(p_1, p_2, \ldots, p_n).$$

$$(1.81)$$

The polarization vectors $\varepsilon_{\pm}^{\mu}$ expressed in terms of the helicity spinors take the form

$$\varepsilon_{+,i}^{\alpha\dot{\alpha}} = -\sqrt{2}\frac{\tilde{\lambda}_i^{\dot{\alpha}}\mu_i^{\alpha}}{\langle\lambda_i\mu_i\rangle}, \qquad \varepsilon_{-,i}^{\alpha\dot{\alpha}} = \sqrt{2}\frac{\lambda_i^{\alpha}\tilde{\mu}_i^{\dot{\alpha}}}{[\lambda_i\mu_i]}, \qquad (1.82)$$

where μ_i and $\tilde{\mu}_i$ are *arbitrary* reference spinors. By acting with the helicity operator
of Eq. (1.75) one easily sees that the $\varepsilon_{\pm,i}$ carry helicity ± 1,

$$h\varepsilon_{\pm,i}^{\alpha\dot{\alpha}} = (\pm 1)\varepsilon_{\pm,i}^{\alpha\dot{\alpha}}. \qquad (1.83)$$

Moreover the properties of Eq. (1.62) are fulfilled

$$(\varepsilon_+)^* = \varepsilon_-,$$

$$k \cdot \varepsilon_{\pm} = \frac{1}{2}\lambda_{\alpha}\tilde{\lambda}_{\dot{\alpha}}\varepsilon_{\pm}^{\alpha\dot{\alpha}} \sim \left([\lambda\lambda] \text{ or } \langle\lambda\lambda\rangle\right) = 0,$$

$$\varepsilon_+ \cdot (\varepsilon_+)^* = \varepsilon_+ \cdot \varepsilon_- = -\frac{2}{2} \frac{\tilde{\lambda}^{\dot\alpha} \mu^\alpha \lambda_\alpha \tilde{\mu}_{\dot\alpha}}{\langle \lambda\mu \rangle [\lambda\mu]} = -1,$$

$$\varepsilon_+ \cdot (\varepsilon_-)^* = \varepsilon_+ \cdot \varepsilon_+ = \frac{2}{2} \frac{\tilde{\lambda}^{\dot\alpha} \mu^\alpha \tilde{\lambda}_{\dot\alpha} \mu_\alpha}{\langle \lambda\mu \rangle^2} = 0.$$

The nature of the reference spinors μ_i and $\tilde{\mu}_i$ with associated reference momentum $q_i^{\alpha\dot\alpha} := \mu_i^\alpha \tilde{\mu}_i^{\dot\alpha}$ corresponds to the freedom of performing local gauge transformations. To see this let us consider the infinitesimal change of the polarization $\varepsilon_+^{\alpha\dot\alpha}$ under a shift of $\mu \to \mu + \delta\mu$

$$\delta\varepsilon_+^{\alpha\dot\alpha} = -\sqrt{2}\left(\frac{\tilde{\lambda}^{\dot\alpha}\delta\mu^\alpha}{\langle\lambda\mu\rangle} - \tilde{\lambda}^{\dot\alpha}\mu^\alpha \frac{\langle\lambda\delta\mu\rangle}{\langle\lambda\mu\rangle^2} \right)$$

$$= -\sqrt{2}\frac{\tilde{\lambda}^{\dot\alpha}}{\langle\lambda\mu\rangle^2} \underbrace{\left(\delta\mu^\alpha \langle\lambda\mu\rangle - \mu^{\dot\alpha}\langle\lambda\delta\mu\rangle \right)}_{=\lambda^\alpha \langle\mu\delta\mu\rangle}$$

$$= p^{\alpha\dot\alpha}\left(\sqrt{2}\frac{\langle\delta\mu\mu\rangle}{\langle\lambda\mu\rangle^2} \right) = p^{\alpha\dot\alpha}\theta(p,\mu,\delta\mu), \qquad (1.84)$$

where Schouten's identity was used in step two. We hence see that the change of the polarization vector ε_+^μ can be understood as a gauge transformation. The amplitude is invariant under the shift $\mu \to \mu + \delta\mu$, since

$$\delta\varepsilon_+^\mu(p)\mathscr{A}_{\mu\nu_1\cdots\nu_{n-1}}^{aa_1\cdots a_{n-1}}(p,q_1,\ldots,q_{n-1}) \sim p^\mu \mathscr{A}_{\mu\nu_1\cdots\nu_{n-1}}^{aa_1\cdots a_{n-1}}(p,q_1,\cdots,q_{n-1})$$

$$= \langle p^\mu A_\mu(p)\cdots \rangle = 0 \qquad (1.85)$$

by transversality. Hence the amplitude depends on the set $\{\lambda_i, \tilde{\lambda}_i, h_i = \pm 1\}$ only and we may choose the reference spinors μ_i and $\tilde{\mu}_i$ for *every* leg at our convenience—reflecting the local gauge invariance of the theory.

1.8 Fermion Polarizations

The polarization spinors for Dirac fermions are known to be given by the particle and anti-particle solutions of the Dirac equation $u_\pm(p)$ and $v_\pm(p)$. In the massless limit these degenerate and are related to the helicity spinors λ and $\tilde{\lambda}$ as discussed in Sect. 1.6

$$u_+(p) = v_-(p) = \binom{\lambda_\alpha}{0} =: |p\rangle, \qquad u_-(p) = v_+(p) = \binom{0}{\tilde{\lambda}^{\dot\alpha}} =: |p], \qquad (1.86)$$

for the chiral representation of the Dirac matrices. We then have with $\bar{\psi} = \psi^\dagger \gamma^0$

$$\bar{u}_+(p) = \bar{v}_-(p) = \left(\lambda_\alpha^* 0\right) \begin{pmatrix} 0 & \mathbb{1} \\ \mathbb{1} & 0 \end{pmatrix} = (0\tilde{\lambda}_{\dot\alpha}) =: [p|,$$

$$\bar{u}_-(p) = \bar{v}_+(p) = \left(0\tilde{\lambda}^{\dot\alpha *}\right) \begin{pmatrix} 0 & \mathbb{1} \\ \mathbb{1} & 0 \end{pmatrix} = \left(\lambda^\alpha 0\right) =: \langle p|.$$

(1.87)

One immediately sees that

$$\langle k|\gamma^\mu|p\rangle = 0 \quad \text{and} \quad [k|\gamma^\mu|p] = 0,$$

(1.88)

as well as

$$\langle p|\gamma^\mu|p] = \lambda^\alpha \sigma_{\alpha\dot\alpha}^\mu \tilde{\lambda}^{\dot\alpha} = 2p^\mu, \qquad [p|\gamma^\mu|p\rangle = 2p^\mu.$$

(1.89)

We also note the relations

$$[1|\slashed{q}|2\rangle = \tilde{\lambda}_{1\dot\alpha} q^{\dot\alpha\alpha} \lambda_{2\alpha}, \qquad \langle 2|\slashed{q}|1] = \lambda_2^\alpha q_{\alpha\dot\alpha} \tilde{\lambda}_1^{\dot\alpha},$$

(1.90)

with $\slashed{q} = q_\mu \gamma^\mu$, and q^μ an arbitrary four-vector implying

$$[i|\gamma^\mu|j\rangle = \langle j|\gamma^\mu|i]$$

(1.91)

upon index raising and lowerings in Eq. (1.90).

The polarization degrees of freedom of external massless fermion states are now expressed through the helicity spinors λ and $\tilde{\lambda}$. In our convention of all external momenta being outgoing we then have

$$|p\rangle \,\&\, \langle p| : -\frac{1}{2} \text{ helicity state;} \qquad |p] \,\&\, [p| : +\frac{1}{2} \text{ helicity state.}$$

(1.92)

Hence outgoing quark lines are represented by $|p]$ and $[p|$ states, whereas outgoing anti-quark lines are represented by $|p\rangle$ and $\langle p|$ states. As a consequence of Eq. (1.88), external fermion lines entering a vertex need to have opposite helicities. Concluding this section, we note the Fierz rearrangement formula

$$[i|\gamma^\mu|j\rangle\langle l|\gamma_\mu|k] = 2[ik]\langle lj\rangle.$$

(1.93)

Exercise 1.3 (Fierz Rearrangement)
Prove the Fierz rearrangement formula (1.93).

1.9 Color Decomposition

Next we discuss a central technique to disentangle the color and kinematical degrees of freedom in a gauge theory scattering amplitude. This decomposition will lead us

Fig. 1.2 Color index
dependence in Feynman rules

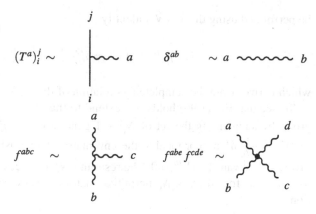

$$(T^a)^j_i \sim \qquad\qquad \delta^{ab} \qquad \sim a \qquad b$$

$$f^{abc} \sim \qquad\qquad f^{abe} f^{cde} \sim$$

to the partial or color-ordered amplitudes, which are independent of color and easier
to construct. In essence we seek a basis spanning the color degrees of freedom of
scattering amplitudes at tree and loop-level.

For this we specialize our discussion from now on to the gauge group $SU(N_c)$,
but keep N_c arbitrary. The color dependent building blocks in an amplitude, as they
derive from the Feynman rules are given in Fig. 1.2.

The color dependence of a generic graph is given by terms involving contractions
of the generator matrices T^a and the structure constants f^{abc}, e.g.

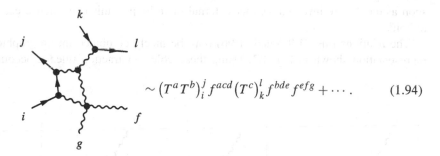

$$\sim (T^a T^b)^j_i \, f^{acd} (T^c)^l_k \, f^{bde} \, f^{efg} + \cdots . \qquad (1.94)$$

Using the key relation Eq. (1.39)

$$f^{abc} = -\frac{i}{\sqrt{2}} \, \mathrm{Tr}\big(T^a[T^b, T^c]\big), \qquad (1.95)$$

for the completely anti-symmetric structure constants, the dependence of a generic
graph on the structure constants f^{abc} may be replaced by products of traces of the
generators T^a, of the form $\mathrm{Tr}(T^a T^{b_1} \cdots) \cdots \mathrm{Tr}(T^a T^{c_1} \cdots)$. External quark lines on
the other hand lead to open strings of the generators T^a such as $(T^{a_1} T^{a_2} \cdots T^{a_n})^j_i$.
Contractions over adjoint indices in these products of strings and traces may then

be performed using the $SU(N_c)$ identity

$$\left(T^a\right)_{i_1}{}^{j_1}\left(T^a\right)_{i_2}{}^{j_2} = \delta_{i_1}^{j_2}\delta_{i_2}^{j_1} - \frac{1}{N_c}\delta_{i_1}^{j_1}\delta_{i_2}^{j_2}, \tag{1.96}$$

which derives from the completeness relation of the $U(N_c)$ gauge group.

To see that Eq. (1.96) holds, we extend to the $U(N_c) = SU(N_c) \times U(1)$ gauge group by augmenting the set of $N_c^2 - 1$ generators $(T^a)_i^j$ by the $U(1)$ generator $(T^0)_i^j := \frac{1}{\sqrt{N_c}}\delta_i^j$ proportional to the unit matrix. Obviously the structure constants involving T^0 vanish: $f^{0ab} = 0$. The resulting N_c^2 matrices T^A with $A = \{0, a\}$ then form a basis for all $N_c \times N_c$ hermitian matrices. They obey a completeness relation

$$\sum_{A=\{0,a\}} \left(T^A\right)_{i_1}^{j_1}\left(T^A\right)_{i_2}^{j_2} = \delta_{i_1}^{j_2}\delta_{i_2}^{j_1}, \tag{1.97}$$

whose normalization is fixed by Eq. (1.38). Moving the $U(1)$ piece to the right-hand-side then yields Eq. (1.96).

Based on this argument we see that the color dependence of any graph may be reduced entirely to traces and strings of the generators $(T^a)_i^j$ with all color adjoint indices corresponding to an external gluon leg and all open fundamental indices corresponding to an external quark line running through the graph. Moreover, for the pure Yang-Mills case the $SU(N_c)$ and the $U(N_c)$ amplitudes are identical as the $U(1)$ 'photon' completely decouples from the non-Abelian $SU(N_c)$ gluons. This is known as the photon decoupling theorem. However, as soon as quarks are involved—either external or in loops—this equivalence ceases to hold.

The relations Eqs. (1.95) and (1.96) may be nicely combined into a graphical representation shown in Fig. 1.3. Using these color contraction rules it becomes

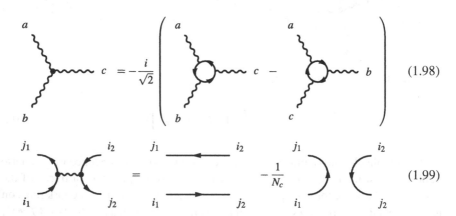

$$\tag{1.98}$$

$$\tag{1.99}$$

Fig. 1.3 Graphical representation of Eqs. (1.95) and (1.96)

Fig. 1.4 Examples of possible poles in a color ordered tree-amplitude

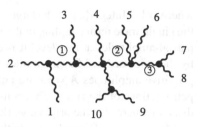

$$① : \quad (p_1 + p_2)^2$$
$$② : \quad (p_9 + p_{10} + p_1 + p_2 + p_3 + p_4)^2$$
$$③ : \quad (p_7 + p_8)^2$$

clear, that the color degrees of freedom of a pure-gluon tree-diagram reduce to a single-trace structure for the generator matrices T^a and may be brought into the *color decomposed* form

$$\mathscr{A}_{n\text{-gluon}}^{\text{tree}}(\{a_i, h_i, p_i\}) = g^{n-2} \sum_{\sigma \in S_n/Z_n} \text{Tr}\left(T^{a_{\sigma_1}} T^{a_{\sigma_2}} \cdots T^{a_{\sigma_n}}\right)$$

$$\times A_{g^n}^{\text{tree}}(p_{\sigma_1}, h_{\sigma_1}; p_{\sigma_2}, h_{\sigma_2}; \ldots; p_{\sigma_n}, h_{\sigma_n}). \quad (1.100)$$

Here the gluon momenta and polarizations are denoted by p_i and $h_i = \pm 1$, while S_n is the set of permutations of n elements and Z_n is the subset of cyclic permutations. Hence S_n/Z_n denotes the set of all non-cyclic permutations of n elements, which is equivalent to S_{n-1}. The sum above runs over all distinct orderings of the external gluon legs. The $A_{g^n}^{\text{tree}}(\{p_i, h_i\})$ are the *partial* or *color-ordered* amplitudes which contain all kinematical information. In fact they are simpler than the full amplitudes, as they only receive contributions from a fixed cyclic ordering of the gluons. Due to this the possible poles of the $A_{g^n}^{\text{tree}}(\{p_i, h_i\})$ may only appear in channels of adjacent momenta $(p_i + p_{i+1} + \cdots + p_{i+s})^2$. An example is given in Fig. 1.4.

At the one-loop order of the perturbative expansion for pure gluon amplitudes it is intuitively clear that more than one trace can survive the reduction process. Indeed the one-loop amplitudes take the color decomposed form (we write compactly $\{p_i, h_i\} := i^{h_i}$)

$$\mathscr{A}_{n\text{-gluon}}^{\text{1-loop}}(\{a_i, h_i, p_i\})$$

$$= g^n \Bigg\{ N_c \sum_{\sigma \in S_n/Z_n} \text{Tr}\left(T^{a_{\sigma_1}} T^{a_{\sigma_2}} \cdots T^{a_{\sigma_n}}\right) A_{g^n;1}\left(\sigma_1^{h_{\sigma_1}}, \ldots, \sigma_n^{h_{\sigma_n}}\right)$$

$$+ \sum_{i=2}^{[n/2]+1} \sum_{\sigma \in S_n/Z_n} \text{Tr}\left(T^{a_{\sigma_1}} \cdots T^{a_{\sigma_{i-1}}}\right) \text{Tr}\left(T^{a_{\sigma_i}} \cdots T^{a_{\sigma_n}}\right) A_{g^n;i}\left(\sigma_1^{h_{\sigma_1}}, \ldots, \sigma_n^{h_{\sigma_n}}\right) \Bigg\},$$

$$(1.101)$$

where $[x] = \text{Integer}[x]$. In the large N_c limit the one-loop amplitude is dominated by the single trace term coupling to the *primitive* amplitude $A_{g^n;1}$. Just as the tree-level partial-amplitudes at tree-level it is color-ordered and may be directly constructed by the color-ordered Feynman rules to be discussed in the next chapter. The higher primitive amplitudes $A_{g^n;i>1}$ are not independent but may be constructed through permutations of the $A_{g^n;1}$. This construction is somewhat involved and will not be discussed here, see the original works [3, 4] for more details.

Moving to QCD amplitudes with one fermion line one has the following color decomposed form at tree-level

$$\mathscr{A}^{\text{tree}}_{g^{n-2}q\bar{q}} = g^{n-2} \sum_{\sigma \in S_{n-2}} \left(T^{a_{\sigma_1}} T^{a_{\sigma_2}} \cdots T^{a_{\sigma_{n-2}}}\right)_i^j$$

$$\times A^{\text{tree}}_{g^{n-2}q\bar{q}}(q_1, h_{q_1}; q_2, h_{q_2}; p_{\sigma_1}, h_{\sigma_1}; \ldots; p_{\sigma_{n-2}}, h_{\sigma_{n-2}}).$$

$$(1.102)$$

Note that in this color basis the quark and anti-quark indices are fixed and do not participate in the permutation sum.

The color structure of $m > 1$ fermion line tree-amplitudes involving $n = 2m$ quarks and anti-quarks may be decomposed into m products of strings in T-matrices, $(T^{a_1} T^{a_2} \cdots T^{a_r})_{i_s}^{j_s}$, where the adjoint indices are either contracted with an outer gluon leg or connect to another fermion line. Using Eq. (1.96) for these internal contractions leads to a final basis of m products of open T-matrix strings with only external adjoint indices. The general construction of this color-decomposition is rather involved and we shall not discuss it here, see [5] or the more recent discussions [6–8] for an explanation. In this case, some of the color factors also include explicit factors of $1/N_c$ deriving from the second term in Eq. (1.96). Nevertheless, all of the kinematical dependences can still be constructed from suitable linear combinations of the partial amplitudes for external quarks and anti-quarks and external gluons generated by the color ordered Feynman diagrams. Hence the partial amplitudes are the atoms of gauge theory scattering amplitudes.

1.10 Color Ordered Feynman Rules and Properties of Color-Ordered Amplitudes

As all color dependences have been removed for the partial amplitudes one may define color-ordered Feynman rules which are free of color degrees of freedom.

Table 1.4 Color-ordered Feynman rules in a compact notation using dummy polarization vectors ε_i^μ and dummy spinors λ_i respectively $\tilde{\lambda}_i$ to absorb the index structures. All momenta are outgoing, outgoing quark lines are represented by $|p]$ and $[p|$ states, whereas outgoing anti-quark lines are represented by $|p\rangle$ and $\langle p|$ states, compare the discussion in Sect. 1.8

Propagators	

$$-\frac{i}{p^2+i0}\eta_{\mu\nu}$$

$$\frac{i\not{p}}{p^2+i0}\eta_{\mu\nu}$$

Color ordered vertices	

$$-g\frac{i}{\sqrt{2}}[(\varepsilon_1\cdot\varepsilon_2)(p_{12}\cdot\varepsilon_3)+(\varepsilon_2\cdot\varepsilon_3)(p_{23}\cdot\varepsilon_1)+(\varepsilon_3\cdot\varepsilon_1)(p_{31}\cdot\varepsilon_2)]$$
$$p_{ij}:=p_i-p_j$$

$$g^2\frac{i}{2}[2(\varepsilon_1\cdot\varepsilon_3)(\varepsilon_2\cdot\varepsilon_4)-(\varepsilon_1\cdot\varepsilon_2)(\varepsilon_3\cdot\varepsilon_4)-(\varepsilon_1\cdot\varepsilon_4)(\varepsilon_2\cdot\varepsilon_3)]$$

$$+g\frac{i}{\sqrt{2}}[3|\not{\varepsilon}_2|1\rangle$$

$$-g\frac{i}{\sqrt{2}}[1|\not{\varepsilon}_2|3\rangle$$

To compute a tree-partial amplitude or a loop-level color-ordered partial amplitude (such as $A_{g^n;1}$) one needs to draw all color-ordered planar graphs reflecting the cyclic ordering of the external legs with no leg-crossings allowed. The resulting graph is then evaluated using the vertices reported in Table 1.4 which are the Feynman rules of Table 1.2 with color dependences removed. The color-ordered amplitudes enjoy general properties which reduce the number of independent amplitudes considerably:

- Cyclicity:

$$A(1, 2, \ldots, n) = A(2, \ldots, n, 1). \tag{1.103}$$

This is an immediate consequence of the definition of the color ordered amplitudes.

- Parity:

$$A(1, 2, \ldots, n) = A(\bar{1}, \bar{2}, \ldots, \bar{n}). \tag{1.104}$$

Where $\bar{i}$ denotes the inversion of the helicity of leg i. Performing this on all legs does not change the amplitude.

- Charge conjugation:

$$A(1_q, 2_{\bar{q}}, 3, \ldots, n) = -A(1_{\bar{q}}, 2_q, 3, \ldots, n). \tag{1.105}$$

Flipping the helicity on a quark line changes the amplitude up to a sign. This is obvious from the color-ordered quark-anti-quark-gluon vertex in Table 1.4.

- Reflection:

$$A^{\text{tree}}(1, 2, \ldots, n) = (-)^n A^{\text{tree}}(n, n-1, \ldots, 1). \tag{1.106}$$

This follows from the anti-symmetry of the color-ordered Feynman rules of Table 1.4 under reflection of all legs, it also holds in the presence of quark lines but only for tree-amplitudes.

- Photon decoupling:

$$A_{g^n}^{\text{tree}}(1, 2, 3, \ldots, n) + A_{g^n}^{\text{tree}}(2, 1, 3, \ldots, n) + A_{g^n}^{\text{tree}}(2, 3, 1, \ldots, n)$$
$$+ \cdots + A_{g^n}^{\text{tree}}(2, 3, \ldots, n-1, 1, n) = 0. \tag{1.107}$$

This relation for pure gluon trees follows from Eq. (1.100): Pure gluon amplitudes in the $U(N)$ theory containing a $U(1)$ photon must vanish. Enforcing this for Eq. (1.100) and collecting all common color-trace terms of identical structure yields the above relation.

- Gauge invariance:
  The color-ordered amplitudes are gauge invariant.

Exercise 1.4 (Color Ordered Four-Point Vertex)
    Derive the form of the color ordered four-point vertex in Table 1.4 from the Feynman rules of Table 1.2.

Exercise 1.5 (Independent 4 and 5 Gluon Partial Amplitudes)
    Use the above relations amongst the color-ordered amplitudes to determine the independent set of color-ordered amplitudes for 4 and 5 gluon scattering.

1.11 Vanishing Tree Amplitudes

Let us now begin with the actual evaluation of the simplest tree-amplitudes, the vanishing ones! Our freedom to choose an arbitrary light-like reference momentum $q_i^{\dot{\alpha}\alpha} = \mu_i^\alpha \tilde{\mu}_i^{\dot{\alpha}}$ in the definition of the gluon polarization vectors $\varepsilon_{\pm,i}$ in Eq. (1.82) may be used to prove the vanishing of certain classes of pure gluon amplitudes.

From Eq. (1.82) we note the polarization vector products

$$\varepsilon_{+,i} \cdot \varepsilon_{+,j} = \frac{\langle \mu_i \mu_j \rangle [\lambda_j \lambda_i]}{\langle \lambda_i \mu_i \rangle \langle \lambda_j \mu_j \rangle},$$

$$\varepsilon_{+,i} \cdot \varepsilon_{-,j} = -\frac{\langle \mu_i \lambda_j \rangle [\mu_j \lambda_i]}{\langle \lambda_i \mu_i \rangle [\lambda_j \mu_j]}, \qquad (1.108)$$

$$\varepsilon_{-,i} \cdot \varepsilon_{-,j} = \frac{\langle \lambda_i \lambda_j \rangle [\mu_j \mu_i]}{[\lambda_i \mu_i][\lambda_j \mu_j]},$$

with the only restriction on the reference spinors being distinct, i.e $\mu_i \neq \lambda_i$ and $\tilde{\mu}_i \neq \tilde{\lambda}_i$, in order to have an independent reference. Making the uniform choice $q_1 = q_2 = \cdots = q_n = q$ with an arbitrary light-like $q \neq p_i$ immediately yields the relations

$$\varepsilon_{+,i} \cdot \varepsilon_{+,j} = 0 = \varepsilon_{-,i} \cdot \varepsilon_{-,j} \quad \forall i, j. \qquad (1.109)$$

An n-gluon tree-amplitude must depend on the n-polarization vectors involved, which have to be contracted with themselves (as $\varepsilon_{\mp,i} \cdot \varepsilon_{\pm,j}$) or with the external momenta (as $p_i \cdot \varepsilon_{\pm,j}$).

What is the minimal number of polarization vector contractions, $\varepsilon_i \cdot \varepsilon_j$, arising in the terms that constitute an n-gluon tree-amplitude? To find this number, we need to look at graphs which maximize the number of momentum-polarization contraction, $p_i \cdot \varepsilon_j$. This implies looking at graphs built entirely out of three-point vertices. Pure three-point vertex n-gluon trees, however, are made of $(n-2)$-vertices. This is seen by induction: the statement is manifestly true for an $n = 3$ gluon tree amplitude. The inductive step attaches a three-vertex to one of the n-legs of the amplitude with—by assumption—$(n-2)$-vertices. This clearly increases also the number of legs by one, which proves the statement by induction.

As an n-leg graph contains n distinct polarization vectors, we conclude that any n-gluon amplitude will consist of terms containing *at least* one polarization contraction $\varepsilon_i \cdot \varepsilon_j$. Armed with this insight, we now prove the vanishing of three classes of tree-amplitudes.

(ii) Choosing the reference momenta q_i of an n-gluon tree-amplitude uniformly as in Eq. (1.109) implies

$$A_n^{\text{tree}}(1^+, 2^+, \ldots, n^+) = 0, \qquad (1.110)$$

as at least one $\varepsilon_{+,i} \cdot \varepsilon_{+,j}$ contraction must arise.

(ii) Similarly, the gluon tree-amplitude with one flipped helicity state vanishes

$$A_n^{\text{tree}}\left(1^-, 2^+, \dots, n^+\right) = 0. \tag{1.111}$$

This follows from the reference momenta choice

$$q_1 = q \neq p_1 \quad \text{and} \quad q_2 = \dots = q_n = p_1, \tag{1.112}$$

as then all terms containing a $\varepsilon_{+,i} \cdot \varepsilon_{+,j} = 0$ contraction with $i, j \in \{2, \dots, n\}$ vanish, and $\varepsilon_{+,i} \cdot \varepsilon_{-,1} = 0$ by the specific choice above.

(iii) Finally, the $q\bar{q}g^{n-2}$ amplitudes vanish, if all gluons have identical helicity

$$A_n^{\text{tree}}\left(1_{\bar{q}}^-, 2_q^+, 3^+, \dots, n^+\right) = 0. \tag{1.113}$$

Now due to the presence of a quark line there is at least one contraction of the form

$$[2|\not{\varepsilon}_{+,i}|1\rangle = \tilde{\lambda}_{2\dot{\alpha}}\varepsilon_{+,i}^{\dot{\alpha}\alpha}\lambda_{1\alpha} = -\sqrt{2}\frac{[\lambda_2\lambda_i]\langle\mu_i\lambda_1\rangle}{\langle\lambda_i\mu_i\rangle} \tag{1.114}$$

in every term constituting the amplitude. Choosing the gluon-polarization reference momenta uniformly as $q_i = \mu_i\tilde{\mu}_i = \lambda_1\tilde{\lambda}_1$ for all $i \in \{3, \dots, n\}$ yields $[2|\not{\varepsilon}_{+,i}|1\rangle = 0$ and hence the vanishing of Eq. (1.113).

In fact we want to restrict to the case of $n > 3$ here, as there are subtleties for the three point amplitudes to be discussed later.

By parity the vanishing of Eqs. (1.110) and (1.111) implies

$$A_n^{\text{tree}}\left(1^\pm, 2^-, \dots, n^-\right) = 0. \tag{1.115}$$

Hence the first non-trivial pure gluon tree-amplitudes are the ones with two flipped helicities $A_n^{\text{tree}}(1^-, 2^+, \dots, (i-1)^+, i^-, (i+1)^+, \dots, n^+)$ known as *maximally helicity violating* (MHV) amplitudes. To understand this name recall that in our convention all momenta are out-going. So by crossing symmetry the MHV amplitudes describes, for example, a process in which all incoming gluons have one helicity and all but two outgoing gluons—the maximal allowed number—have the opposite helicity: Helicity is not conserved and this process is maximally helicity violating.

Similarly we have for the single-quark-line-gluon amplitude

$$A_n^{\text{tree}}\left(1_{\bar{q}}^-, 2_q^+, 3^- \dots, n^-\right) = 0, \tag{1.116}$$

by choosing $q_i = \mu_i\tilde{\mu}_i = \lambda_2\tilde{\lambda}_2$ for all $i \in \{3, \dots, n\}$. Alternatively this may be seen by a parity and charge conjugation transformation of Eq. (1.113).

We will see in Sect. 2.7.3 that the vanishing of the amplitudes discussed above can be understood from a hidden supersymmetry in quark-gluon tree-amplitudes.

Fig. 1.5 Color-ordered
graphs contributing to the
four-gluon MHV amplitude

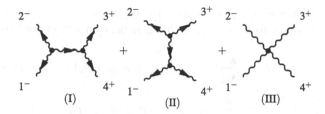

1.12 The Four-Gluon Tree-Amplitude

We shall now compute the simplest non-trivial gluon amplitude, namely the four-gluon MHV-amplitude with a split helicity distribution

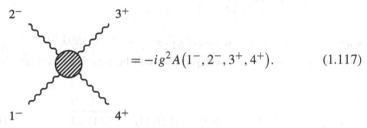

$$= -ig^2 A\left(1^-, 2^-, 3^+, 4^+\right). \qquad (1.117)$$

Employing the color-ordered Feynman rules of Table 1.4 we see that the three diagrams of Fig. 1.5 contribute to the amplitude. Its computation is again considerably simplified by a clever choice of the reference momenta $q_i = \mu_i \tilde{\mu}_i$ of the gluon-polarizations

$$q_1 = q_2 = p_4, \qquad q_3 = q_4 = p_1, \qquad (1.118)$$

where p_i denote the physical external momenta. Then one sees, using Eq. (1.108), that the polarization vector products

$$\varepsilon_{-,1} \cdot \varepsilon_{-,2} = \varepsilon_{+,3} \cdot \varepsilon_{+,4} = \varepsilon_{-,1} \cdot \varepsilon_{+,4} = \varepsilon_{-,2} \cdot \varepsilon_{+,4} = \varepsilon_{-,1} \cdot \varepsilon_{+,3} = 0 \qquad (1.119)$$

all vanish and the *only* non-vanishing contraction is

$$\varepsilon_{-,2} \cdot \varepsilon_{+,3} = -\frac{\langle \mu_3 \lambda_2 \rangle [\mu_2 \lambda_3]}{\langle \lambda_3 \mu_3 \rangle [\lambda_2 \mu_2]} = -\frac{\langle 12 \rangle [34]}{\langle 13 \rangle [24]}. \qquad (1.120)$$

Of course it is mandatory to use the same choice of reference momenta for all graphs.

Diagram I:

$$
\begin{aligned}
&= \left(-\frac{ig}{\sqrt{2}}\right)^2 \frac{-i\eta_{\mu\nu}}{s_{12}} \left[\varepsilon^\mu_{-,2}(p_{2q} \cdot \varepsilon_{-,1}) + \varepsilon^\mu_{-,1}(p_{q1} \cdot \varepsilon_{-,2}) \right] \\
&\quad \times \left[\varepsilon^\nu_{+,4}(p_{4(-q)} \cdot \varepsilon_{+,3}) + \varepsilon^\nu_{+,3}(p_{(-q)3} \cdot \varepsilon_{+,4}) \right],
\end{aligned}
$$

$$(1.121)$$

with $q = -p_1 - p_2 = p_3 + p_4$, $s_{12} = (p_1 + p_2)^2 = \langle 12\rangle[21]$ and $p_{ij} = p_i - p_j$. Due to only $\varepsilon_{-,2} \cdot \varepsilon_{+,3}$ surviving in the contraction, we have

$$\begin{aligned}
(I) &= \frac{ig^2}{2s_{12}}(\varepsilon_{-,2} \cdot \varepsilon_{+,3})(p_2 + p_1 + p_2) \cdot \varepsilon_{-,1}(-p_3 - p_4 - p_3) \cdot \varepsilon_{+,4} \\
&= -\frac{2ig^2}{s_{12}}(\varepsilon_{-,2} \cdot \varepsilon_{+,3})(p_2 \cdot \varepsilon_{-,1})(p_3 \cdot \varepsilon_{+,4}) \\
&= -i\frac{2ig^2}{s_{12}}\left(-\frac{\langle 12\rangle[34]}{\langle 13\rangle[24]}\right)\left(\frac{1}{\sqrt{2}}\frac{\langle 12\rangle[24]}{[14]}\right)\left(\frac{1}{\sqrt{2}}\frac{\langle 13\rangle[34]}{[41]}\right) \\
&= -ig^2\frac{\langle 12\rangle[34]^2}{[12]\langle 14\rangle[41]},
\end{aligned} \tag{1.122}$$

where in the second line we used $p_2 \cdot \varepsilon_{-,1} = \frac{1}{\sqrt{2}}\frac{\langle 12\rangle[2\mu_1]}{[1\mu_1]} \overset{\mu_1=4}{=} \frac{1}{\sqrt{2}}\frac{\langle 12\rangle[24]}{[14]}$ and similarly for $p_3 \cdot \varepsilon_{+,4}$. The last expression can be rewritten exclusively in λ_i spinors as

$$\begin{aligned}
\frac{i}{g^2}(I) &= \frac{\langle 12\rangle[34]^2}{[12]\langle 14\rangle[41]} \cdot \frac{\langle 43\rangle}{\langle 43\rangle} = \frac{\langle 12\rangle[34]}{[12]\langle 14\rangle}\frac{\overbrace{[34]\langle 43\rangle}^{\langle 12\rangle[21]}}{\underbrace{\langle 43\rangle[41]}_{-\langle 34\rangle[41]=\langle 32\rangle[21]}} = -\frac{\langle 12\rangle^2}{\langle 23\rangle\langle 41\rangle}\underbrace{\frac{[34]}{[21]}}_{\frac{\langle 12\rangle}{\langle 43\rangle}} \\
&= \frac{\langle 12\rangle^3}{\langle 23\rangle\langle 34\rangle\langle 41\rangle}.
\end{aligned} \tag{1.123}$$

Here in the first step the four-point kinematical relation $s_{12} = \langle 12\rangle[21] = s_{34} = \langle 34\rangle[43]$ was used.

Diagram II:

Using again the fact that the only non-vanishing contraction is Eq. (1.120) we find a vanishing result

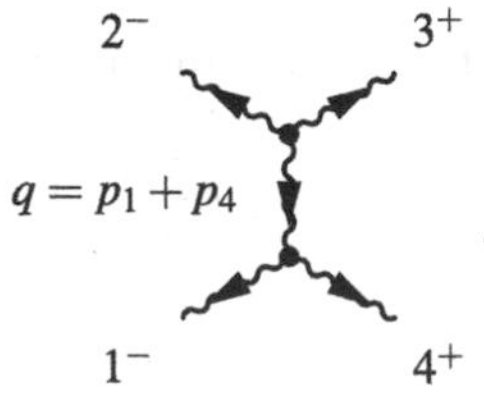

$$\begin{aligned}
&\sim \frac{-i}{2s_{12}}(\varepsilon_2 \cdot \varepsilon_3)\big[(p_{23} \cdot \varepsilon_1)(p_{1(-q)} \cdot \varepsilon_4) \\
&\quad + (p_{23} \cdot \varepsilon_4)(p_{(-q)4} \cdot \varepsilon_1)\big] \\
&= 0.
\end{aligned} \tag{1.124}$$

Here with $q = p_1 + p_4$ we have $p_{1(-q)} \cdot \varepsilon_4 = 2p_1 \cdot \varepsilon_4 = 0$ which vanishes by virtue of our choice $q_4 = p_1$ of Eq. (1.118). Similarly $p_{(-q)4} \cdot \varepsilon = -2p_4 \cdot \varepsilon_1 = 0$ as $q_1 = p_4$. Hence diagram II gives no contribution.

Diagram III:

The same holds true for the contact graph as

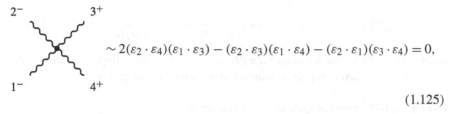

$$\sim 2(\varepsilon_2 \cdot \varepsilon_4)(\varepsilon_1 \cdot \varepsilon_3) - (\varepsilon_2 \cdot \varepsilon_3)(\varepsilon_1 \cdot \varepsilon_4) - (\varepsilon_2 \cdot \varepsilon_1)(\varepsilon_3 \cdot \varepsilon_4) = 0,$$

(1.125)

where always at least one of the vanishing contractions in Eq. (1.119) appears.

In summary we have established the compact result

$$A(1^-, 2^-, 3^+, 4^+) = \frac{\langle 12 \rangle^4}{\langle 12 \rangle \langle 23 \rangle \langle 34 \rangle \langle 41 \rangle}.$$

(1.126)

At this point it is also instructive to check the helicities of the result individually for every leg using the local helicity generator of Eq. (1.75)

$$h_i = \frac{1}{2}\left(-\lambda_i^\alpha \frac{\partial}{\partial \lambda_i^\alpha} + \tilde{\lambda}_i^{\dot{\alpha}} \frac{\partial}{\partial \tilde{\lambda}_i^{\dot{\alpha}}}\right),$$

(1.127)

to wit

$$h_1 A = \frac{1}{2}(-4+2)A = -A, \qquad h_2 A = -A,$$

$$h_3 A = +A, \qquad h_4 A = +A.$$

(1.128)

By cyclicity there is only one more independent four-gluon amplitude left to compute: the case $A(1^-, 2^+, 3^-, 4^+)$ with an alternating helicity distribution. All other possible helicity distributions can be related to this or $A(1^-, 2^-, 3^+, 4^+)$ of Eq. (1.126) by cyclicity. It turns out, that we do not need to do another Feynman diagrammatic computation, as the missing amplitude follows from the $U(1)$ decoupling theorem of Sect. 1.10:

$$A(1^-, 2^+, 3^-, 4^+) = -A(1^-, 2^+, 4^+, 3^-) - A(1^-, 4^+, 2^+, 3^-)$$

$$= -\left(\frac{\langle 31 \rangle^4}{\langle 12 \rangle \langle 24 \rangle \langle 43 \rangle \langle 31 \rangle} + \frac{\langle 31 \rangle^4}{\langle 14 \rangle \langle 42 \rangle \langle 23 \rangle \langle 31 \rangle}\right)$$

$$= \frac{\langle 31 \rangle^4}{\langle 12 \rangle \langle 23 \rangle \langle 34 \rangle \langle 41 \rangle}.$$

(1.129)

Comparing this result with Eq. (1.126) we may express all four-gluon MHV amplitudes in a single and handy formula

$$A_{g^4}^{\text{tree}}(\ldots, i^-, \ldots, j^-, \ldots) = \frac{\langle ij \rangle^4}{\langle 12 \rangle \langle 23 \rangle \langle 34 \rangle \langle 41 \rangle},$$

(1.130)

where the dots stand for positive helicity gluon states. In fact we shall show in Chap. 2 that this formula has a straightforward generalization to the n-gluon MHV case in which only the denominator is modified to

$$A_{g^n}^{\text{tree}}(\ldots, i^-, \ldots, j^-, \ldots) = \frac{\langle ij \rangle^4}{\langle 12 \rangle \langle 23 \rangle \cdots \langle (n-1)n \rangle \langle n1 \rangle}, \tag{1.131}$$

known as the Parke-Taylor amplitude [9]. It is a remarkably simple and closed expression for tree-level amplitude with arbitrary number n of gluons.

Exercise 1.6 (Non-trivial Quark-Gluon Scattering)

Show by using the color ordered Feynman rules and a suitable choice for the gluon-polarization reference vector $q_3^{\alpha\dot\alpha} = \mu_3^\alpha \tilde\mu_3^{\dot\alpha}$ that the first non-trivial $\bar q q g g$ scattering amplitude is given by

$$A_{\bar q q g^2}^{\text{tree}}(1_{\bar q}^-, 2_q^+, 3^-, 4^+) = \frac{\langle 13 \rangle^3 \langle 23 \rangle}{\langle 12 \rangle \langle 23 \rangle \langle 34 \rangle \langle 41 \rangle}. \tag{1.132}$$

Also convince yourself that the result for this amplitude has the correct helicity assignments as was done in Eq. (1.128) for the pure-gluon case.

1.13 References and Further Reading

The first five sections of this chapter cover material discussed in standard textbooks on quantum field theory. In particular the more recent textbook [10] also covers the material of the later sections of this chapter such as spinor helicities, color decomposition and colored ordered Feynman rules.

The use of commuting spinors for helicity amplitudes goes back to the 1980s and was introduced by a number of authors [11–14] who realized their utility for the compact description of massless tree-amplitudes. The general MHV amplitude was proposed in [9] and proved in [15]. The method of color decomposition for scattering amplitudes has its roots in open string scattering amplitudes and was introduced in [5]. At the one-loop order the color decomposition of gauge theory amplitudes was first discussed in [3, 4]. More recent general accounts of color decomposition techniques are [6–8, 16].

References

1. C.-N. Yang, R.L. Mills, Conservation of isotopic spin and isotopic gauge invariance. Phys. Rev. **96**, 191 (1954)
2. G. 't Hooft, Some twisted self dual solutions for the Yang-Mills equations on a hypertorus. Commun. Math. Phys. **81**, 267–275 (1981)
3. Z. Bern, D.A. Kosower, Color decomposition of one loop amplitudes in gauge theories. Nucl. Phys. B **362**, 389 (1991)

4. Z. Bern, L.J. Dixon, D.A. Kosower, One loop corrections to two quark three gluon amplitudes. Nucl. Phys. B **437**, 259 (1995). arXiv:hep-ph/9409393

5. M.L. Mangano, S.J. Parke, Multi-parton amplitudes in gauge theories. Phys. Rep. **200**, 301–367 (1991). arXiv:hep-th/0509223

6. H. Ita, K. Ozeren, Colour decompositions of multi-quark one-loop QCD amplitudes. J. High Energy Phys. **1202**, 118 (2012). arXiv:1111.4193

7. S. Badger, B. Biedermann, P. Uwer, V. Yundin, Numerical evaluation of virtual corrections to multi-jet production in massless QCD. Comput. Phys. Commun. **184**, 1981–1998 (2013). arXiv:1209.0100

8. C. Reuschle, S. Weinzierl, Decomposition of one-loop QCD amplitudes into primitive amplitudes based on shuffle relations (2013). arXiv:1310.0413

9. S.J. Parke, T.R. Taylor, An amplitude for n gluon scattering. Phys. Rev. Lett. **56**, 2459 (1986)

10. M. Srednicki, *Quantum Field Theory* (Cambridge University Press, Cambridge, 2007)

11. P. De Causmaecker, R. Gastmans, W. Troost, T.T. Wu, Multiple bremsstrahlung in gauge theories at high-energies. 1. General formalism for quantum electrodynamics. Nucl. Phys. B **206**, 53 (1982)

12. F.A. Berends, R. Kleiss, P. De Causmaecker, R. Gastmans, W. Troost, et al., Multiple bremsstrahlung in gauge theories at high-energies. 2. Single bremsstrahlung. Nucl. Phys. B **206**, 61 (1982)

13. R. Kleiss, W.J. Stirling, Spinor techniques for calculating $p\bar{p} \to W_\pm / Z_0 +$ jets. Nucl. Phys. B **262**, 235–262 (1985)

14. Z. Xu, D.-H. Zhang, L. Chang, Helicity amplitudes for multiple bremsstrahlung in massless nonabelian gauge theories. Nucl. Phys. B **291**, 392 (1987)

15. F.A. Berends, W.T. Giele, Recursive calculations for processes with n gluons. Nucl. Phys. B **306**, 759 (1988)

16. T. Schuster, Color ordering in QCD (2013). arXiv:1311.6296

References

Chapter 2
Tree-Level Techniques

In this chapter, we discuss techniques to compute tree-level scattering amplitudes. Feynman rules allow one to write down any tree-level amplitude as a sum of all Feynman diagrams contributing, as we did in the last chapter for the four-gluon amplitude. In practice, this straightforward approach can quickly become cumbersome, as the number of Feynman diagrams typically grows factorially with the number of external legs. To give an example: for the scattering of two gluons into n gluons the number of contributing Feynman diagrams grows from 4 ($n = 2$), to 220 ($n = 4$), to 34,300 ($n = 6$) to an enormous 10,525,900 for $n = 8$ [1]. Moreover, one often finds that the final answer is much simpler than the intermediate steps of the calculation.

Here, we will present techniques that exploit the reasons behind this simplicity. Very powerful insights can be gained from thinking about tree-level amplitudes as algebraic functions of the external momenta. As we will see, their analyticity properties under complex deformations of the momenta can be used to derive simple, yet powerful recursion relations known as Britto-Cachazo-Feng-Witten (BCFW) on-shell recursions. These recursion relations use as input on-shell amplitudes, so that the gauge redundancy, which is partly responsible for the complexity of conventional Feynman graph calculations, is absent.

We will also discuss in some detail the symmetry properties of tree-level scattering amplitudes, that is Poincaré symmetry and in the massless case the extension to conformal symmetry. We then briefly introduce the concept of supersymmetry and present the maximally supersymmetric gauge theory in four dimensions, namely $\mathcal{N} = 4$ super Yang-Mills theory, which plays a distinguished role in the space of all gauge theories. We discuss its scattering amplitudes, a supersymmetric variant of the BCFW recursion and the extended superconformal symmetry, and discuss how results for non-supersymmetric theories can be obtained from the supersymmetric case.

2.1 Britto-Cachazo-Feng-Witten (BCFW) On-shell Recursion

The Britto-Cachazo-Feng-Witten (BCFW) recursion relations are an efficient way to compute higher-point partial amplitudes from lower point ones. As we will see,

J.M. Henn, J.C. Plefka, *Scattering Amplitudes in Gauge Theories*,
Lecture Notes in Physics 883, DOI 10.1007/978-3-642-54022-6_2,
© Springer-Verlag Berlin Heidelberg 2014

the mere knowledge of three-point amplitudes allows the construction of *all* higher point amplitudes in a recursive fashion.

To begin with let us consider an n-gluon amplitude $A_n(p_1, \ldots, p_n)$. The key to the BCFW recursion is to study how the amplitude behaves under a complex deformation of the particle momenta preserving the on-shell conditions. For this we perform the following complex shift of the helicity spinors for two neighboring legs 1 and n,

$$
\begin{aligned}
\lambda_1 &\to \hat{\lambda}_1(z) = \lambda_1 - z\lambda_n, \\
\tilde{\lambda}_n &\to \hat{\tilde{\lambda}}_n(z) = \tilde{\lambda}_n + z\tilde{\lambda}_1,
\end{aligned}
\tag{2.1}
$$

with $z \in \mathbb{C}$, and $\tilde{\lambda}_1$ and λ_n are left inert. We denote the shifted quantities by a hat. This corresponds to a complexification of momenta

$$
\begin{aligned}
p_1^{\alpha\dot\alpha} &\to \hat{p}_1^{\alpha\dot\alpha}(z) = (\lambda_1 - z\lambda_n)^\alpha \tilde{\lambda}_1^{\dot\alpha}, \\
p_2^{\alpha\dot\alpha} &\to \hat{p}_2^{\alpha\dot\alpha}(z) = \lambda_n^\alpha (\tilde{\lambda}_n + z\tilde{\lambda}_1)^{\dot\alpha}.
\end{aligned}
\tag{2.2}
$$

Importantly, the deformation preserves both overall momentum conservation and the on-shell conditions,

$$
\hat{p}_1^2(z) = 0, \qquad \hat{p}_n^2(z) = 0, \qquad \hat{p}_1(z) + \hat{p}_n(z) = p_1 + p_2,
\tag{2.3}
$$

and the z-deformed partial amplitude may be written as

$$
\mathscr{A}_n(z) = \delta^{(4)}\left(\sum_{i=1}^n p_i\right) A_n(z).
\tag{2.4}
$$

The BCFW recursion relations rely on an understanding of the behavior of the function $A_n(z)$ in the complex z plane. The derivation proceeds in three steps. First, the locations of the poles of $A_n(z)$ are analyzed. Then, it is shown that the residues of the poles correspond to products of lower-point tree amplitudes. Finally, the large z behavior of $A_n(z)$ is determined.

What are the analytical properties in z of the deformed amplitude $A_n(z)$? In order to answer that question, it is helpful to think of $A_n(z)$ in terms of tree-level Feynman diagrams that contribute to it. The sum of Feynman diagrams is gauge invariant. Therefore we can choose the Feynman gauge for the following discussion, without loss of generality. It is clear that $A_n(z)$ is a rational function of the $\lambda_i, \tilde{\lambda}_i$ and z. Moreover, $A_n(z=0)$ can only have poles where the denominators of Feynman propagators become zero. Given that the partial amplitudes are color-ordered, the propagators take the form

$$
\frac{1}{(p_i + p_{i+1} + \cdots + p_j)^2},
\tag{2.5}
$$

i.e. they are given by a sum of adjacent momenta. Cf. Fig. 1.4 for illustration. It follows that the deformed amplitude $A_n(z)$ will only have simple poles in z of the

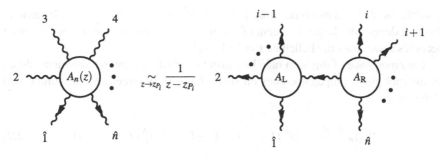

Fig. 2.1 The factorization of the z deformed amplitude on the pole $z = z_{P_i}$

form

$$\frac{1}{\hat{P}_i(z)} := \frac{1}{(\hat{p}_1(z) + p_2 + \cdots + p_{i-1})^2} = \frac{1}{(p_i + p_{i+1} + \cdots + \hat{p}_n(z))^2}$$

$$= \frac{1}{P_i^2 - z\langle n|P_i|1]}, \tag{2.6}$$

with the region momenta $P_i := p_1 + p_2 + \cdots + p_{i-1}$ and $\langle n|P_i|1] = \lambda_{n\alpha} P_i^{\alpha\dot{\alpha}} \tilde{\lambda}_{1\dot{\alpha}}$. Note that any region momentum containing both $\hat{p}_1$ and $\hat{p}_n$ is independent of z, and hence cannot contribute any poles. Therefore the only propagators that can produce poles are the ones considered above.

We deduce that $A_n(z)$ has *simple poles* in z at positions

$$z_{P_i} = \frac{P_i^2}{\langle n|P_i|1]}, \quad \forall i \in [3, n-1]. \tag{2.7}$$

We also need to know what the residues at the poles are. In fact tree-level amplitudes have universal factorization properties when propagators go on-shell. This is easy to see from Feynman diagrams. The on-shell propagator splits the Feynman diagrams into two parts, or clusters. One can convince oneself that each cluster contains all Feynman diagrams that would be required to compute a scattering amplitude. Put differently the propagator going on-shell can be represented as inserting a complete set of all on-shell states. In conclusion, near the pole z_{P_i} the amplitude $A_n(z)$ factorizes into a product of lower-point amplitudes, which we refer to as "left" and "right" amplitudes A^L and A^R, respectively (see Fig. 2.1),

$$\lim_{z \to z_{P_i}} A_n(z) = \frac{1}{z - z_{P_i}} \frac{-1}{\langle n|P_i|1]} \sum_s A^L\big(\hat{1}(z_{P_i}), 2, \ldots, i-1, -\hat{P}^s(z_{P_i})\big)$$

$$\times A^R\big(\hat{P}^{\bar{s}}(z_{P_i}), i, \ldots, n-1, \hat{n}(z_{P_i})\big). \tag{2.8}$$

The key point is that the internal propagator goes on-shell, so that one has a product of on-shell subamplitudes A^L and A^R. The sum over s in Eq. (2.8) runs over all possible on-shell states propagating between A^L and A^R and $\bar{s} = -s$. In general this will depend on the field content of the theory under consideration. For gluons it becomes a sum over the helicities $s = \{+1, -1\}$.

The emergence of this sum may be justified with the following argument. When an internal gluon propagator goes on-shell, $P^2 = 0$, the tree-level amplitude will factorize as

$$P^2 A_n \xrightarrow{P^2=0} -i M_\mu^L(1, \ldots, i-1, -P)\eta^{\mu\nu} M_\nu^R(P, i, \ldots, n), \qquad (2.9)$$

where $M_\mu^{L/R}$ are lower-point amplitudes with the polarization vector on the $\pm P$ leg stripped off. Without loss of generality let us parametrize the on-shell momentum four-vector as $P^\mu = E(1, 0, 0, 1)$ and the gluon polarization vectors as

$$\varepsilon_\pm^\mu = \frac{1}{\sqrt{2}}(0, 1, \pm i, 0). \qquad (2.10)$$

A straightforward calculation yields

$$\sum_{s=+,-} e_s^\mu M_\mu^L e_{\bar{s}}^\mu M_\mu^R = M_1^L M_1^R + M_2^L M_2^R, \qquad (2.11)$$

while the transversality of the amplitude $P^\mu M_\mu^{L/R} = 0$ implies $M_0^{L/R} - M_3^{L/R} = 0$. Adding this zero to the above Eq. (2.11) yields

$$\begin{aligned}
-i M_\mu^L \eta^{\mu\nu} M_\nu^R &= i\left(M_1^L M_1^R + M_2^L M_2^R + M_3^L M_3^R - M_0^L M_0^R\right) \\
&= i \sum_{s=+,-} e_s^\mu M_\mu^L e_{\bar{s}}^\mu M_\mu^R \\
&= i \sum_{s=+,-} A^L\left(1, \ldots, i-1, -P^s\right) A^R\left(P^{\bar{s}}, i, \ldots, n\right), \qquad (2.12)
\end{aligned}$$

as stated in Eq. (2.8). We note that the same argument would also go through if we were to put an internal fermion propagator on-shell. Then the numerator of the "cut" propagator reads $(|P\rangle[P| + |P]\langle P|)$, which immediately introduces the sum over fermion helicities $\pm 1/2$. For a scalar propagator put on-shell, finally, there is no helicity sum and Eq. (2.8) is obvious.

Returning to our discussion, we can now use complex analysis to construct $A_n(z = 0)$ from the knowledge of the poles of $A_n(z)$. Consider the function $A_n(z)/z$. It behaves as

$$\lim_{z \to z_{P_i}} \frac{A_n(z)}{z} = -\frac{1}{z - z_{P_i}} \sum_s A_L^s(z_{P_i}) \frac{1}{P_i^2} A_R^{\bar{s}}(z_{P_i}), \qquad (2.13)$$

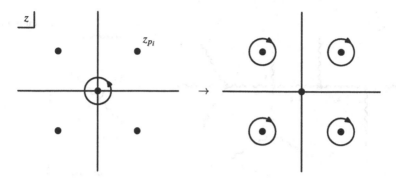

Fig. 2.2 Pulling the initial circle C_0 off to infinity in Eq. (2.16)

with the abbreviations

$$A_L^s(z_{P_i}) = A_i^L\big(\hat{1}(z_{P_i}), 2, \ldots, i-1, -\hat{P}^s(z_{P_i})\big), \tag{2.14}$$

$$A_R^{\bar{s}}(z_{P_i}) = A_i^R\big(\hat{P}^{\bar{s}}(z_{P_i}), i, \ldots, n-1, \hat{n}(z_{P_i})\big). \tag{2.15}$$

Of course we are interested in the original amplitude $A_n = A_n(z = 0)$. Using the residue theorem it may be written as

$$A_n = A_n(z=0) = \oint_{C_0} \frac{dz}{2\pi i} \frac{A(z)}{z}$$

$$= \sum_{i=2}^{n-1} \sum_s A_L^s(z_{P_i}) \frac{1}{P_i^2} A_R^{\bar{s}}(z_{P_i}) + \text{Res}(z = \infty). \tag{2.16}$$

Here C_0 is a small circle around the origin at $z = 0$ not embracing any of the poles z_{P_i}. To reach the final expression we have pulled this circle off to infinity capturing all the poles z_{P_i} in the complex plane, now encircled in the opposite orientation, see Fig. 2.2. If $A_n(z) \to 0$ as $z \to \infty$ we can drop the residue at $z = \infty$ (also called boundary term). As we show presently, this is the case for gauge theories, under certain conditions. Assuming this for now we arrive at the BCFW recursion relation [2]:

$$A_n = \sum_{i=2}^{n-1} \sum_s A_L^s(z_{P_i}) \frac{1}{P_i^2} A_R^{\bar{s}}(z_{P_i}). \tag{2.17}$$

The following comments are in order. We chose neighboring legs $\hat{1}$ and $\hat{n}$ to perform the complex shifts. One may generalize to non-adjacent legs. In that case, there are typically more BCFW diagrams to consider. In general, different BCFW deformations lead to equivalent representations of the same amplitude. The equivalence may not always be easy to see analytically. Multi-line shifts involving more than two legs have also been considered in the literature [3].

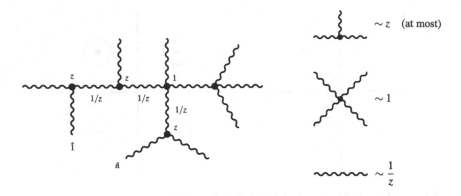

Fig. 2.3 Maximal z scaling of an individual Feynman diagram contributing to $A_n(z)$

In order to complete the derivation of the BCFW recursion (2.17), we need to determine the large z behavior of $A_n(z)$. To have $\oint_\infty \frac{dz}{2\pi i} \frac{A_n(z)}{z} = 0$ we need that $A_n(z)$ vanishes as $z \to \infty$.

In order to estimate the large z behavior of generic tree-level amplitudes we perform an analysis based on Feynman graphs. There are three sources for z dependence: the denominators of the propagators, the interaction vertices, and the polarization vectors.

Consider a generic graph contributing to the tree-level n-gluon amplitude ($\hat{1}$ and $\hat{n}$ are assumed to be neighbors). As one can see from Fig. 2.3 the z dependence occurs only along the path from $\hat{1}$ to $\hat{n}$. Along this path, each three-gluon vertex, being linear in the momenta, contributes a factor of z at most, four-gluon vertices contribute 1, and propagators behave as $1/z$.

We can derive an upper bound by considering the least favorable case. This is when the line $\hat{1}$ to $\hat{n}$ contains only three-valent vertices. In that case it is easy to see that the graph behaves as $\sim z$.

Finally, there is an additional z-dependence arising from polarization vectors at legs 1 and n:

$$\varepsilon_1^{+\alpha\dot\alpha} = -\sqrt{2}\frac{\tilde{\lambda}_1^{\dot\alpha}\mu_1^\alpha}{\langle\hat\lambda_1(z)\mu_1\rangle} \sim \frac{1}{z}, \qquad \varepsilon_1^{-\alpha\dot\alpha} = \sqrt{2}\frac{\hat\lambda_1^\alpha(z)\tilde\mu_1^{\dot\alpha}}{[\lambda_1\mu_1]} \sim z, \qquad (2.18)$$

$$\varepsilon_n^{+\alpha\dot\alpha} = -\sqrt{2}\frac{\hat{\tilde\lambda}_n^{\dot\alpha}(z)\mu_n^\alpha}{\langle\lambda_n\mu_n\rangle} \sim z, \qquad \varepsilon_n^{-\alpha\dot\alpha} = \sqrt{2}\frac{\lambda_n^\alpha\tilde\mu_n^{\dot\alpha}}{[\hat\lambda_n(z)\mu_n]} \sim \frac{1}{z}. \qquad (2.19)$$

Therefore, taking all sources of z dependence into account, we conclude that individual graphs scale at worst as

$$A(\hat{1}^+\hat{n}^-) \sim \frac{1}{z}, \qquad A(\hat{1}^+\hat{n}^+) \sim z, \qquad (2.20)$$

$$A(\hat{1}^-\hat{n}^-) \sim z, \qquad A(\hat{1}^-\hat{n}^+) \sim z^3. \qquad (2.21)$$

This shows that the $(+-)$-shift has the desired falloff properties that allow to drop the boundary term at infinity in the BCFW formula. By cyclicity it is always possible to find a $\{\hat{1}^+, \hat{n}^-\}$ pair. In fact, the above bound is too conservative. One can show [4] that the $(++)$ and $(--)$-shifts also lead to an overall $\frac{1}{z}$ scaling once the sum over all Feynman graphs is performed, as the terms scaling as z or 1 cancel out. Only the $(-+)$-shift gives a non-vanishing $\mathrm{Res}(z = \infty)$ in general.

So far we have discussed the recursion for pure gluon amplitudes. In fact the arguments used to reach Eq. (2.16) go through for an arbitrary quantum field theory. However, the vanishing of the residue at infinity may not. For example one might consider a seemingly simple scalar ϕ^4-theory. Here it is easy to convince oneself that the residue at infinity will not vanish. For two neighboring shifted legs in the ϕ^4 there are always diagrams contributing to the amplitude in which $\hat{1}$ and $\hat{n}$ are connected via a single ϕ^4-vertex. As the external legs have no polarization degrees of freedom such a Feynman diagram will not depend on z. There is no chance of a cancellation either, as all other diagrams in which $\hat{1}$ and $\hat{n}$ are connected by more than one vertex vanish $z \to \infty$. Thus the amplitude $A_n(z)$ for a scalar field theory will not vanish for $z \to \infty$. In a sense the apparent simplicity of ϕ^4 theory at the level of the Feynman rules is misleading, as this theory does not give rise to a simple on-shell recursion relation. The ϕ^4-theory is often used as a testing ground for quantum field theory. Here we see a first hint that, from the on-shell perspective, Yang-Mills theory is actually simpler.

Example As an example of the general result above, consider for concreteness the four-gluon amplitude derived in Eq. (1.130) Under a $(--)$-shift, we have

$$A_4\big(\hat{1}^-, 2^+, 3^+, \hat{4}^-\big) = \frac{\langle \hat{1}\hat{4}\rangle^4}{\langle \hat{1}2\rangle\langle 23\rangle\langle 3\hat{4}\rangle\langle \hat{4}\hat{1}\rangle} = \frac{\langle 14\rangle^4}{\langle \hat{1}2\rangle\langle 23\rangle\langle 34\rangle\langle 41\rangle} \sim \frac{1}{z}, \quad (2.22)$$

where we have used

$$\langle \hat{4}\hat{1}\rangle \overset{|\hat{4}\rangle = |4\rangle}{=} \langle 4\hat{1}\rangle = \langle 41\rangle - z\langle 44\rangle = \langle 41\rangle. \quad (2.23)$$

This is in line with the claim that the actual z-scaling of the $(--)$ and $(++)$ shifts is better than estimated in Eq. (2.20). On the other hand, under a $(-+)$ shift, the amplitude behaves as

$$A_4\big(\hat{1}^-, 2^-, 3^+, \hat{4}^+\big) = \frac{\langle \hat{1}2\rangle^4}{\langle \hat{1}2\rangle\langle 23\rangle\langle 34\rangle\langle 41\rangle} \sim z^3. \quad (2.24)$$

This is consistent with the general bounds derived above.

We will now proceed by explaining an important subtlety concerning three-point amplitudes. This will complete all prerequisites for using the BCFW recursion relations, and we will then derive the n-point MHV amplitudes as a first example.

2.2 The Gluon Three-Point Amplitude

In order to use the BCFW recursion relations derived above we need to understand the starting point of the recursion, in other words the 'smallest' or atomistic amplitudes. It may be slightly surprising that these are certain three-particle amplitudes. As we discuss presently, their definition is somewhat peculiar due to the constraints imposed by the on-shell conditions and momentum conservation.

In fact, for real null momenta there are no three-point amplitudes, since

$$p_1^\mu + p_2^\mu + p_3^\mu = 0 \quad \text{implies} \quad p_1 \cdot p_2 = p_2 \cdot p_3 = p_3 \cdot p_1 = 0. \tag{2.25}$$

All Mandelstam invariants vanish and there are no other Lorentz scalars an amplitude could depend on

$$p_i \cdot p_j = 0 \quad \leftrightarrow \quad \langle ij \rangle [ji] = 0. \tag{2.26}$$

The situation is different for *complex momenta* $p_i \in \mathbb{C}$. In this case the helicity spinors λ_i and $\tilde{\lambda}_i$ are independent, and the conditions $p_i \cdot p_j = 0$ can be solved either by $\langle ij \rangle = 0 \; \forall i, j = 1, 2, 3$ or by $[ij] = 0 \; \forall i, j = 1, 2, 3$. Hence either $\lambda_1^\alpha \propto \lambda_2^\alpha \propto \lambda_3^\alpha$ (collinear left-handed spinors) or $\tilde{\lambda}_1^\alpha \propto \tilde{\lambda}_2^\alpha \propto \tilde{\lambda}_3^\alpha$ (collinear right-handed spinors) solve the constraints $p_i \cdot p_j = 0$. The two choices correspond to the three-gluon MHV_3 and $\overline{\mathrm{MHV}}_3$ amplitudes, respectively. They are given by

$$A_3^{\mathrm{MHV}}(i^-, j^-) = \frac{\langle ij \rangle^4}{\langle 12 \rangle \langle 23 \rangle \langle 31 \rangle},$$
$$[12] = [23] = [31] = 0, \tag{2.27}$$

and

$$A_3^{\overline{\mathrm{MHV}}}(i^+, j^+) = -\frac{[ij]^4}{[12][23][31]},$$
$$\langle 12 \rangle = \langle 23 \rangle = \langle 31 \rangle = 0, \tag{2.28}$$

respectively. These three-point amplitudes can of course be straightforwardly read off from the color-ordered Feynman rules. Alternatively, one can argue that these are the only functional forms compatible with the helicity assignments of the external particles and the vanishing of the $[ij]$ and $\langle ij \rangle$, respectively, up to a free constant which is identified with the coupling constant [5]. Let us illustrate this point for the MHV_3 amplitude. Without loss of generality we can take the negative helicity states to be $i = 1, j = 2$. From the kinematical discussion above it is clear that the amplitude can depend only on the variables $\langle 12 \rangle$, $\langle 23 \rangle$ and $\langle 31 \rangle$. The helicity assignment implies

$$h_{1,2} A_3^{\mathrm{MHV}} = -A_3^{\mathrm{MHV}}, \qquad h_3 A_3^{\mathrm{MHV}} = A_3^{\mathrm{MHV}}, \tag{2.29}$$

where we recall that the helicity generator was defined as

$$h_i = -\frac{1}{2} \lambda_i^\alpha \frac{\partial}{\partial \lambda_i^\alpha} + \frac{1}{2} \tilde{\lambda}_i^{\dot{\alpha}} \frac{\partial}{\partial \tilde{\lambda}_i^{\dot{\alpha}}}. \tag{2.30}$$

These equations fix the answer to be $\langle 12 \rangle^3 / (\langle 23 \rangle \langle 31 \rangle)$, up to the overall normalization.

In summary, we have arrived at a remarkable result. Via the BCFW recursion we can produce all n-gluon tree amplitudes from the three-point amplitude. The structure of the latter follows solely from kinematical considerations, i.e. helicity assignments and momentum conservation. In particular, the explicit form of the four-gluon vertex in Yang-Mills theory is not needed!

Exercise 2.1 (Three-Point Amplitudes from Color-Ordered Feynman Rules)

Verify the expressions for the three-point MHV amplitude of Eq. (2.27) and the three-point anti-MHV amplitude of Eq. (2.28) by an explicit calculation using the color-ordered Feynman rules of Chap. 1, given in Table 1.4.

2.3 An Example: MHV Amplitudes

Having established the BCFW recursion relations, we will now give a simple yet non-trivial example, and derive the formula for n-point MHV amplitudes with arbitrary multiplicity n.

For simplicity, but without loss of generality, let us assume that the negative helicity gluons are at positions 1 and n. Then, by the discussion above we will obtain a valid BCFW relation without boundary term if we shift the momenta p_1 and p_n, as in Eq. (2.1), as this corresponds to a $(--)$-shift.

We then have to consider all BCFW diagrams that can contribute to this channel, see Fig. 2.4. There are two possible helicity assignments for the internal propagator, $(-+)$ and $(+-)$. As we will see presently, this imposes further constraints on the BCFW diagrams. In the $(+-)$ case, A_L has only one negative helicity gluon. As we saw in Sect. 1.12, such an amplitude vanishes unless it has three external legs. This leads us to draw the first BCFW diagram shown in Fig. 2.4. Likewise, the $(-+)$ helicity assignment leads to a zero term unless only three legs enter A_R. However, there is an additional subtlety. In fact, with the $\overline{\mathrm{MHV}}_3$ amplitude on the right the three-point condition $\hat{\tilde{\lambda}}_n \propto \lambda_{n-1}$ implies $\langle \hat{n} n - 1 \rangle = \langle n n - 1 \rangle = 0$, i.e. the collinearity of p_n and p_{n-1}. For generic external momenta, which is the case we are

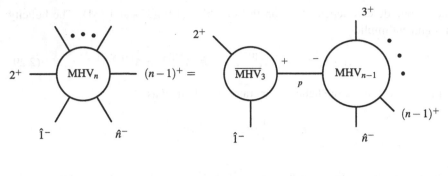

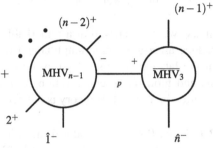

Fig. 2.4 The BCFW recursion for an MHV$_n$ amplitude

interested in, this collinearity is not fulfilled and therefore this contribution vanishes, since the amplitude on the right does. Therefore, we are left with only one BCFW diagram, the first one on the RHS of Fig. 2.4.

Let us now prove the formula for MHV amplitudes by induction. We want to prove that

$$A_n^{\mathrm{MHV}}\left(n^-, 1^-\right) = \frac{\langle n1 \rangle^4}{\langle 12 \rangle \cdots \langle n1 \rangle}. \tag{2.31}$$

We already know that this formula is true for $n = 3$, see Eq. (2.27) (and also for $n = 4$, from the Feynman graph calculation of Sect. 1.12). Therefore we only need to prove the inductive step.

We will also need the formula for the three-point $\overline{\mathrm{MHV}}$ amplitude that was given in Eq. (2.28). In the BCFW channel that we are considering, we have

$$z_P = \frac{P^2}{\langle n|P|1]}, \qquad P^\mu = p_1^\mu + p_2^\mu, \tag{2.32}$$

and hence

$$z_P = \frac{(p_1 + p_2)^2}{\langle n|P|1]} = \frac{\langle 12 \rangle [21]}{\langle n2 \rangle [21]} = \frac{\langle 12 \rangle}{\langle n2 \rangle}. \tag{2.33}$$

The amplitudes A_L and A_R are given by the induction assumption

$$A_L = A_3^{\overline{\text{MHV}}}(\hat{1}^-, 2^+, -\hat{P}^+) = -\frac{[2(-\hat{P})]^3}{[12][(-\hat{P})1]}, \tag{2.34}$$

$$A_R = A_{n-1}^{\text{MHV}}(\hat{P}^-, 3^+, 4^+, \ldots, (n-1)^+, \hat{n}^-) = \frac{\langle \hat{n}\hat{P}^-\rangle^3}{\langle \hat{P}3\rangle \langle 34\rangle \cdots \langle (n-1)\hat{n}\rangle}. \tag{2.35}$$

Then, using $|-\tilde{\lambda}_P] = -|\tilde{\lambda}_P]$ and $|-\lambda_P\rangle = +|\lambda_P\rangle$, and

$$[\hat{1}*] = [1*], \qquad \langle \hat{n}*\rangle = \langle n*\rangle, \qquad \langle n\hat{P}\rangle[\hat{P}2] = \langle n\hat{1}\rangle[12] = \langle n1\rangle[12], \tag{2.36}$$

$$(p_1 + p_2)^2 = \langle 12\rangle[21], \qquad \langle 3\hat{P}\rangle[\hat{P}1] = \langle 32\rangle[2\hat{1}] = \langle 32\rangle[21], \tag{2.37}$$

we find

$$\hat{A}_L \frac{1}{(p_1 + p_2)^2} \hat{A}_R = -\frac{\langle n1\rangle^3[12]^3}{[12][21]\langle 32\rangle[21]\langle 12\rangle\langle 34\rangle \cdots \langle (n-1)n\rangle} = \frac{\langle n1\rangle^4}{\langle 12\rangle \cdots \langle n1\rangle}, \tag{2.38}$$

in perfect agreement with Eq. (2.31). This completes the inductive proof of Eq. (2.31).

Exercise 2.2 (The 6-Gluon Split-Helicity NMHV Amplitude)
Determine the first non-trivial next-to-maximally-helicity-violating (NMHV) amplitude

$$A_6^{\text{tree}}(1^+, 2^+, 3^+, 4^-, 5^-, 6^-)$$

from the BCFW recursion relation and our knowledge of the MHV amplitudes. Consider a shift of the two helicity states 1^+ and 6^- and show that

$$A_6^{\text{tree}}(1^+, 2^+, 3^+, 4^-, 5^-, 6^-) = \frac{\langle 6|p_{12}|3]^3}{\langle 61\rangle\langle 12\rangle[34][45]\langle 5|p_{16}|2]} \frac{1}{(p_6 + p_1 + p_2)^2}$$

$$+ \frac{\langle 4|p_{56}|1]^3}{\langle 23\rangle\langle 34\rangle[16][65]\langle 5|p_{16}|2]} \frac{1}{(p_5 + p_6 + p_1)^2},$$

where $p_{ij} = p_i + p_j$.

2.4 Factorization Properties of Tree-Level Amplitudes

As we have already seen in the derivation of the BCFW recursion relations, the analytic structure of tree-level amplitudes is governed by propagator poles and their residues. The residues are given by lower-point tree-level amplitudes. Here we summarize these properties.

2.4.1 Factorization on Multi-Particle Poles

Partial or color-ordered amplitudes can have poles when region momenta $P_{i,j} :=$ $p_i + p_{i+1} + \cdots + p_j$ go on shell. Then the amplitude factorizes according to

$$A_n^{\text{tree}}(1, \ldots, n) \sim \sum_\lambda A_L\big(i, \ldots, j, P^\lambda\big) \frac{1}{P_{i,j}^2} A_R\big(P^{-\lambda}, j+1, \ldots, i-1\big). \quad (2.39)$$

We call this a two-particle pole if the region momentum $P_{i,j}$ is formed by two external momenta, and multi-particle pole otherwise.

2.4.2 Absence of Multi-Particle Poles in MHV Amplitudes

Multi-gluon amplitudes will in general have multi-particle poles. However, MHV amplitudes are special, and in fact, they only have two-particle poles. The reason is the following. Consider the general factorization formula Eq. (2.39). In a factorization of an MHV amplitude there are only three negative helicity legs (corresponding to the two external negative helicities, and one for the internal on-shell propagator) that are distributed over two partial amplitudes. However, we saw in the previous chapter that $A_n(1^\pm, 2^+, \ldots, n^+) = 0$ (for $n > 3$). Therefore, this is always zero unless one partial amplitude is a three-particle amplitude, and this corresponds to a two-particle pole. In fact we have seen this principle at work in the previous Sect. 2.3. Here the BCFW recursion for MHV_n amplitudes reduced to a single term with a $\overline{\text{MHV}}_3$ amplitude on the left-hand-side, as a consequence of the absence of multi-particle poles.

2.4.3 Collinear Limits

A special case of the factorization formula (2.39) occurs for $j = i+1$, when we have a two-particle singularity. In fact, since the factorization involves a three-particle amplitude, such a pole can only occur for collinear external momenta. We already know from Sect. 2.2 that this is very subtle. For real momenta, the limiting configuration is $p_i \sim p_{i+1}$, which we may parameterize by $p_i = zP$ and $p_{i+1} = (1-z)P$ with the total collinear momentum $P = p_i + p_{i+1}$. It is convenient to write the null-momenta p_i and p_{i+1} in terms of spinors $P = \lambda_P \tilde{\lambda}_P$ associated to P. We then have $\lambda_i = \sqrt{z}\lambda_P$, $\tilde{\lambda}_i = \sqrt{z}\tilde{\lambda}_P$, and similarly for p_{i+1} with z replaced by $1 - z$. If in addition $z \to 1$ (or $z \to 0$), we have a *soft limit*, where the four-momentum of one of the external momenta goes to zero. Tree (and loop-level) amplitudes possess important factorization properties with universal features in collinear as well as in soft limits. We will discuss the tree-level case here (the same analysis also applies to loop integrands).

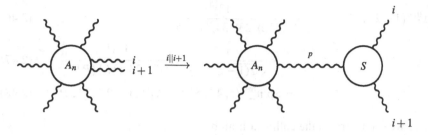

Fig. 2.5 Factorization in the collinear limit of legs i and $i + 1$ where S denotes the splitting function $\mathrm{Split}^{\mathrm{tree}}_{-\lambda}(z, i, i + 1)$

An analysis based on Feynman diagrams shows that tree-level amplitudes have a universal (singular) behavior in the collinear limit. It is governed by splitting functions,

$$A_n^{\mathrm{tree}}\left(\dots, i^{\lambda_i}, (i + 1)^{\lambda_{i+1}}, \dots\right)$$

$$\xrightarrow{i\|i+1} \sum_{\lambda=\pm} \mathrm{Split}^{\mathrm{tree}}_{-\lambda}(z, i, i + 1) A_{n-1}^{\mathrm{tree}}\left(\dots, P^\lambda, \dots\right), \tag{2.40}$$

see Fig. 2.5. The splitting amplitude depends on the helicities of the collinear gluons but is independent of the helicities of the other legs not participating in the collinear limit. This is known as the universality of the splitting functions. We have (see [6] and references therein)

$$\mathrm{Split}^{\mathrm{tree}}_-\left(z, a^-, b^-\right) = 0, \tag{2.41}$$

$$\mathrm{Split}^{\mathrm{tree}}_-\left(z, a^+, b^+\right) = \frac{1}{\sqrt{z(1-z)}\langle ab \rangle}, \tag{2.42}$$

$$\mathrm{Split}^{\mathrm{tree}}_-\left(z, a^+, b^-\right) = -\frac{z^2}{\sqrt{z(1-z)}[ab]}, \tag{2.43}$$

$$\mathrm{Split}^{\mathrm{tree}}_-\left(z, a^-, b^+\right) = -\frac{(1-z)^2}{\sqrt{z(1-z)}[ab]}. \tag{2.44}$$

The remaining splitting amplitudes may be obtained from the ones above by parity,

$$\mathrm{Split}^{\mathrm{tree}}_{-(-\lambda)}\left(z, a^{-\lambda_a}, b^{-\lambda_b}\right) = -\mathrm{Split}^{\mathrm{tree}}_{-\lambda}\left(z, a^{\lambda_a}, b^{\lambda_b}\right)\Big|_{\langle ab \rangle \leftrightarrow [ab]}. \tag{2.45}$$

We shall now derive these expressions from the MHV amplitudes.

2.4.3.1 Example: Collinear Limits of the Five-Point MHV Amplitude

Let us test the above splitting functions using the example of a five-point MHV amplitude,

$$A_5^{\text{tree}}\left(1^-, 2^-, 3^+, 4^+, 5^+\right) = \frac{\langle 12 \rangle^4}{\langle 12 \rangle \langle 23 \rangle \langle 34 \rangle \langle 45 \rangle \langle 51 \rangle} \tag{2.46}$$

$$\xrightarrow{4\|5} \frac{1}{\sqrt{z(1-z)}\langle 45 \rangle} \times \frac{\langle 12 \rangle^4}{\langle 12 \rangle \langle 23 \rangle \langle 3P \rangle \langle P1 \rangle} \tag{2.47}$$

$$= \text{Split}_-^{\text{tree}}\left(z, 4^+, 5^+\right) \times A_4^{\text{tree}}\left(1^-, 2^-, 3^+, P^+\right), \tag{2.48}$$

where we parametrized the collinear limit by

$$\lambda_4 = \sqrt{z}\lambda_P, \qquad \lambda_5 = \sqrt{1-z}\lambda_P, \tag{2.49}$$

$$\tilde{\lambda}_4 = \sqrt{z}\tilde{\lambda}_P, \qquad \tilde{\lambda}_5 = \sqrt{1-z}\tilde{\lambda}_P. \tag{2.50}$$

Indeed agreement with Eq. (2.42) is found. Similarly, we can take the collinear limit in a $(+-)$ channel. We have

$$A_5^{\text{tree}}\left(1^-, 2^-, 3^+, 4^+, 5^+\right) \xrightarrow{2\|3} \frac{z^2}{\sqrt{z(1-z)}} \frac{1}{\langle 23 \rangle} \frac{\langle 1P \rangle^4}{\langle 1P \rangle \langle P4 \rangle \langle 45 \rangle \langle 51 \rangle} \tag{2.51}$$

$$= \text{Split}_+^{\text{tree}}\left(z, 2^-, 3^+\right) \times A_4^{\text{tree}}\left(1^-, P^-, 4^+, 5^+\right) \tag{2.52}$$

from which we deduce

$$\text{Split}_+^{\text{tree}}\left(z, a^-, b^+\right) = \frac{z^2}{\sqrt{z(1-z)}\langle ab \rangle}. \tag{2.53}$$

Converting this to $\text{Split}_-^{\text{tree}}(z, a^+, b^-)$ via the parity transformation Eq. (2.45) recovers Eq. (2.43). In order to check Eq. (2.44) we consider the collinear limit in a $(-+)$-channel

$$A_5^{\text{tree}}\left(1^-, 2^-, 3^+, 4^+, 5^+\right) \xrightarrow{5\|1} \underbrace{\frac{(1-z)^2}{\sqrt{z(1-z)}\langle 51 \rangle}}_{\text{Split}_+^{\text{tree}}(z, 5^+, 1^-)} A_4^{\text{tree}}\left(P^-, 2^-, 3^+, 4^+\right) \tag{2.54}$$

yielding Eq. (2.44) via parity. In order to check the vanishing of Eq. (2.41) one has to study the collinear factorization of the 6-point MHV amplitude with the helicity distributions $A_6^{\text{tree}}(1^-, 2^-, 3^+, 4^+, 5^+, 6^+)$ along the legs 5 and 6.

2.4.4 Soft Limit

One speaks of the soft limit of an external gluon when its energy vanishes. This is equivalent to sending the on-shell four-momentum k_s of the gluon leg s to zero.

This limit also leads to a factorization and has universal features,

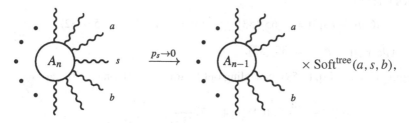

$$A_n^{\text{tree}}(\ldots, a, s^{\pm}, b, \ldots) \xrightarrow{k_s \to 0} \text{Soft}^{\text{tree}}(a, s^{\pm}, b) A_{n-1}^{\text{tree}}(\ldots, a, b, \ldots). \qquad (2.55)$$

The factorized function depends on the momenta and helicities of the soft gluon and the momenta of the color-ordered neighbors a and b. It is independent, however, of the helicities and particle types of the neighboring legs. From considering the soft limit of an MHV amplitude one easily establishes

$$\text{Soft}^{\text{tree}}(a, s^+, b) = \frac{\langle ab \rangle}{\langle as \rangle \langle sb \rangle}. \qquad (2.56)$$

Via parity in analogy to Eq. (2.45) we have

$$\text{Soft}^{\text{tree}}(a, s^-, b) = -\frac{[ab]}{[as][sb]}. \qquad (2.57)$$

The factorization properties under collinear and soft limits may be used to test obtained results for their consistency. For example, the 6-gluon split-helicity amplitude $A_6^{\text{tree}}(1^+, 2^+, 3^+, 4^-, 5^-, 6^{+-})$ has to factorize into a soft function and the 5-point MHV amplitude when its legs 4, 5 or 6 are taken soft. It is instructive to work this out in detail for the leg 4.

In Exercise 2.1 the 6-gluon split-helicity amplitude was computed using the BCFW recursion, with the result

$$A_6^{\text{tree}}(1^+, 2^+, 3^+, 4^-, 5^-, 6^-)$$

$$= \frac{\langle 6|p_{12}|3]^3}{\langle 61 \rangle \langle 12 \rangle [34][45][5|p_{16}|2\rangle} \frac{1}{(p_6 + p_1 + p_2)^2}$$

$$+ \frac{\langle 4|p_{56}|1]^3}{\langle 23 \rangle \langle 34 \rangle [16][65][5|p_{16}|2\rangle} \frac{1}{(p_5 + p_6 + p_1)^2},$$

where $p_{ij} := p_i + p_j$. $\qquad (2.58)$

Taking the soft limit on leg 4, i.e. $|4\rangle \to 0$, we see that the second term in the above formula vanishes, while the first term develops the expected pole of $\text{Soft}^{\text{tree}}(3, 4^-, 5)$ from (2.57)

$$A_6^{\text{tree}} \xrightarrow{4 \to 0} -\frac{[35]}{[34][45]} \frac{\langle 6|p_1 + p_2|3]^3}{\langle 61 \rangle \langle 12 \rangle [53][5|p_1 + p_6|2\rangle} \frac{1}{(p_6 + p_1 + p_2)^2}. \qquad (2.59)$$

Using momentum conservation $p_1 + p_2 + p_3 + p_5 + p_6 = 0$ for the limiting kinematics one has

$$\langle 6|p_1 + p_2|3] = -\langle 65\rangle[53], \qquad [5|p_1 + p_6|2\rangle = -[53]\langle 32\rangle,$$

$$(p_6 + p_1 + p_2)^2 = \langle 35\rangle[53]. \tag{2.60}$$

Plugging this into Eq. (2.59), canceling out factors of [53], one arrives at

$$A_6^{\text{tree}} \xrightarrow{4^- \to 0} \text{Soft}^{\text{tree}}(3, 4^-, 5) \frac{\langle 56\rangle^3}{\langle 12\rangle\langle 23\rangle\langle 35\rangle\langle 61\rangle}$$

$$= \text{Soft}^{\text{tree}}(3, 4^-, 5) A_5^{\text{tree}}(1^+, 2^+, 3^+, 5^-, 6^{+-}), \tag{2.61}$$

which is the correct factorized result of Eq. (2.55).

Exercise 2.3 (The Vanishing Splitting Function $\text{Split}_+^{\text{tree}}(z, a^+, b^+)$)

Show by studying the factorization properties of the six-point MHV-amplitude $A_6^{\text{tree}}(1^-, 2^-, 3^+, 4^+, 5^+, 6^+)$ in the collinear limit $5 \parallel 6$ that

$$\text{Split}_+^{\text{tree}}(z, a^+, b^+) = 0.$$

Exercise 2.4 (Soft Limit of 6-Gluon Split-Helicity Amplitude)

Check the consistency of the 6-gluon split-helicity amplitude of Eq. (2.58) with the soft limit of leg 5. In contrast to the discussion in Sect. 2.4.4, this tests both terms contributing to the amplitude in Eq. (2.58).

Exercise 2.5 (A $\bar{q}qggg$ Amplitude from Collinear and Soft Limits)

In Chap. 1 we established the following color-ordered $\bar{q}qgg$ amplitudes involving a massless quark and anti-quark using color-ordered Feynman rules:

$$A_4^{\text{tree}}\left(1_{\bar{q}}^-, 2_q^+, 3^+, 4^+\right) = 0, \tag{2.62}$$

$$A_4^{\text{tree}}\left(1_{\bar{q}}^-, 2_q^+, 3^-, 4^+\right) = \frac{\langle 13\rangle^3\langle 23\rangle}{\langle 12\rangle\langle 23\rangle\langle 34\rangle\langle 41\rangle}. \tag{2.63}$$

Use these and the discussed splitting and soft factorization properties for gluonic legs to make a sophisticated guess for the five-point single quark-line tree amplitude $A_4^{\text{tree}}(1_{\bar{q}}^-, 2_q^+, 3^-, 4^+, 5^+)$. Check your guess against all known factorization properties.

Can you generalize your guess to the partial amplitudes $A_n^{\text{tree}}(1_{\bar{q}}^-, 2_q^+, 3^+, \ldots, n^+)$ and $A_n^{\text{tree}}(1_{\bar{q}}^-, 2_q^+, 3^-, 4^+, \ldots, n^+)$?

2.5 On-shell Recursion for Amplitudes with Massive Particles

So far our discussion focused on massless amplitudes, mostly involving gluons and massless fermions. Of course amplitudes with massive particles are of great relevance as well and on-shell recursions have also been developed for this case. Let us then briefly discuss tree-level scattering amplitudes of n particles, some of which (but not all) are allowed to be massive

$$A_n(p_1, p_2, \ldots, p_n), \quad p_i^2 = m_i^2. \tag{2.64}$$

Concretely, we consider gauge theory amplitudes involving at least two massless gluons, which we moreover take to be neighboring for simplicity of the discussion, say $p_1^2 = p_n^2 = 0$. Let us now see how the BCFW on-shell recursion derived in Sect. 2.1 may be generalized to gauge theory amplitudes with massive particles. We closely follow reference [7] in our exposition.

As before we consider a complex shift of the null gluon momenta at positions 1 and n by a parameter $z \in \mathbf{C}$

$$\hat{p}_1(z) = p_1 - z|n\rangle[1|,$$

$$\hat{p}_n(z) = p_n + z|n\rangle[1|, \tag{2.65}$$

$$\hat{P}_i(z) = P_i - z|n\rangle[1|,$$

where $P_i = p_1 + \cdots + p_{i-1}$ with $i \in [3, n-1]$ is the sum of adjacent momenta from 1 to $i-1$ which featured in the BCFW recursion relation of Eq. (2.17). In fact, the arguments leading to the BCFW recursion relation are very general and are applicable to a generic quantum field theory. They were obtained by thinking about the poles of the deformed amplitude $A(z)$ as an analytic function in z which arise whenever an internal propagator associated to the momentum $\hat{P}_i(z)$ goes on-shell. This reasoning does not change in the massive case, i.e. when $\hat{P}_i^2(z) = m^2$. The pole then reads in generalization of Eq. (2.6) as

$$\frac{1}{\hat{P}_i(z) - m_{P_i}^2} = \frac{1}{(\hat{p}_1(z) + p_2 + \cdots + p_{i-1})^2 - m_{P_i}^2}$$

$$= \frac{1}{(p_i + p_{i+1} + \cdots + \hat{p}_n(z))^2 - m_{P_i}^2}$$

$$= \frac{1}{P_i^2 - m_{P_i}^2 - z\langle n|P_i|1]}, \tag{2.66}$$

where m_{P_i} is the mass of the associated particle going on-shell. The location of the pole is then simply shifted to

$$z_{P_i} = \frac{P_i^2 - m_{P_i}^2}{\langle n|P_i|1]}, \quad \forall i \in [3, n-1], \tag{2.67}$$

generalizing Eq. (2.7) to the massive case. Using again the theorem that the sum of residues of the rational function $A(z)/z$ on the Riemann sphere is zero, one immediately arrives at the on-shell recursion relation for amplitudes including massive particles

$$A_n(1,\ldots,n) = \sum_{i=2}^{n-1} \sum_s A_L\big(\hat{1}(z_{P_i}), 2, \ldots, i-1, -\hat{P}^s(z_{P_i})\big) \frac{1}{P_i^2 - m_{P_i}^2}$$

$$\times A_R\big(\hat{P}^{\bar{s}}(z_{P_i}), i, \ldots, n-1, \hat{n}(z_{P_i})\big) + \mathrm{Res}(z=\infty). \qquad (2.68)$$

where the sum over s is over all contributing states and the legs 1 and n are massless. Of course this formula is only of use if one can show that the residue at infinity vanishes. This turns out to be the case if the gluon helicities of the shifted legs are not of the $(-,+)$ type as before, i.e. the statement is

$$\mathrm{Res}(z=\infty) = 0 \quad \text{iff} \quad (h_1, h_n) = (+,-), (+,+), (-,-), \qquad (2.69)$$

see [7] for a derivation. This concludes our discussion of the massive on-shell recursion.

Let us now study as an example four point amplitudes involving gluons and massive scalars.

Example (Amplitudes with Gluons and Massive Scalars) We consider a theory of a massive complex scalar field coupled to gauge theory. Concretely, we want to evaluate the four-point amplitude involving two neighboring gluons of positive helicity

$$A_4\big(1^+, 2_\phi, 3_{\bar{\phi}}, 4^+\big). \qquad (2.70)$$

The scalars have mass m^2. In fact this amplitude vanishes in the massless limit $m = 0$, similar to the vanishing of the above amplitude when the scalars are replaced by massless fermions, as was shown in chapter one. This is related to a hidden supersymmetry of massless gauge theory amplitudes to be discussed soon. In fact amplitudes of the above type are of interest even in massless theories at the one-loop level. There the need to regulate divergences leads one to consider internal particles propagating in $D = 4 - 2\varepsilon$ dimensions, as we shall discuss in detail in the next chapter. A massless particle in D dimensions may be alternatively viewed as a massive one in four dimension and hence the above amplitude becomes relevant here. This will be used later on in Sect. 3.5.3.

Returning to our concrete example we employ the massive on-shell recursion of Eq. (2.68). Only a single channel contributes

$$A_4\big(1^+, 2_\phi, 3_{\bar{\phi}}, 4^+\big) = A_3\big(\hat{1}^+, 2_\phi, -\hat{P}_{\bar{\phi}}\big) \frac{1}{P^2 - m^2} A_3\big(\hat{P}_\phi, 3_{\bar{\phi}}, 4^+\big). \qquad (2.71)$$

All that is needed are—again—the atomistic $(\phi g \bar{\phi})$-amplitudes. These follow from the color ordered Feynman vertices of two scalars of mass m and momenta l_1, l_2, and a single gluon with momentum p

$$V_3\big(l_1^+, p^\mu, l_2^-\big) = \frac{i}{\sqrt{2}}\big(l_1^\mu - l_2^\mu\big), \qquad (2.72)$$

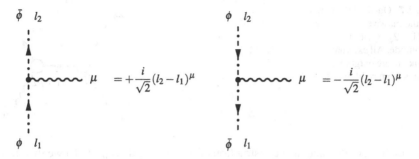

Fig. 2.6 Color ordered Feynman rules for a massive complex scalar field

where the $+$ or $-$ index for the scalars denotes a scalar or anti-scalar state respectively, see Fig. 2.6. Contracting this with the positive helicity gluon polarization of Eq. (1.82) one obtains the on-shell three-point amplitudes

$$A_3\left(l_1^+, p^+, l_2^-\right) = \frac{\langle q|l_1|p]}{\langle qp\rangle} = A_3\left(l_1^-, p^+, l_2^+\right), \tag{2.73}$$

where the last relation follows by reflection. Note that here q is the arbitrary null reference momentum of the gluon leg related to the local gauge invariance of the theory. By similar arguments one establishes the three point amplitudes involving a negative helicity gluon

$$A_3\left(l_1^+, p^-, l_2^-\right) = \frac{\langle p|l_1|q]}{[pq]} = A_3\left(l_1^-, p^-, l_2^+\right). \tag{2.74}$$

An alert reader might object at this point that there appears to be a serious problem with these amplitudes, as they explicitly depend on the reference momentum q. How should one be able to build all scattering amplitudes in this theory on such an arbitrariness? In fact, the amplitudes of Eqs. (2.73) and (2.74) are independent of the choice of q. This is seen as follows. Taking $|q\rangle$ and $|p\rangle$ as a basis in Weyl spinor space we may parametrize an arbitrary reference spinor different from $|q\rangle$ as $|q'\rangle = \alpha|q\rangle + \beta|p\rangle$. Clearly Eq. (2.73) is invariant under rescalings of the reference spinor, thus without loss of generality we may parameterize $|q'\rangle = |q\rangle + \gamma|p\rangle$ or infinitesimally $\delta_q|q\rangle \propto |p\rangle$. This entails that the amplitude Eq. (2.73) changes under a variation of the reference spinor as

$$\delta_q A_3\left(l_1^+, p^+, l_2^-\right) \propto \frac{\langle p|l_1|p]}{\langle qp\rangle} = 0, \tag{2.75}$$

where the vanishing follows from $\langle p|l_1|p] = 2p \cdot l_1 = 0$, which is a consequence of the three point kinematics

$$(l_1 + p)^2 = l_2^2 \quad \rightarrow \quad l_1 \cdot p = 0 \quad \text{as } l_1^2 = l_2^2 = m^2, \ p^2 = 0. \tag{2.76}$$

Fig. 2.7 On-shell recursion
for the massive
$A_4(1^+, 2_\phi, 3_{\bar\phi}, 4^+)$
amplitude. All external
momenta are outgoing,
P runs from right to left

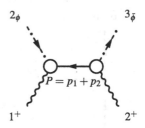

Coming back to the recursive construction of the amplitude Eq. (2.70) we then have
(cf. Fig. 2.7)

$$A_4\left(1^+, 2_\phi, 3_{\bar\phi}, 4^+\right) = A_3\left(-\hat{P}_{\bar\phi}, \hat{1}^+, 2_\phi\right)\frac{1}{P^2 - m^2}A_3\left(3_{\bar\phi}, \hat{4}^+, \hat{P}_\phi\right)$$

$$= -\frac{\langle q_1|\hat{P}|\hat{1}]}{\langle q_1\hat{1}\rangle}\frac{1}{P^2 - m^2}\frac{\langle q_4|p_3|\hat{4}]}{\langle q_4\hat{4}\rangle}. \tag{2.77}$$

Here $q_{1/4}$ denote the reference momenta of the gluon legs 1 and 4. Things are simplified considerably with the gauge choice

$$q_1 = \hat{p}_4, \qquad q_4 = \hat{p}_1. \tag{2.78}$$

Noting that $|\hat{1}] = |1]$ and $|\hat{4}\rangle = |4\rangle$ we then have

$$\langle q_1|\hat{P}|\hat{1}] = \langle 4|\hat{P}|1] = \langle 4|P|1] = -\langle 4|p_3|1], \qquad \langle q_1\hat{1}\rangle = \langle 4\hat{1}\rangle = \langle 41\rangle,$$
$$\langle q_4|p_3|\hat{4}] = \langle\hat{1}|p_3|\hat{4}], \qquad \langle q_4\hat{4}\rangle = \langle\hat{1}4\rangle = \langle 14\rangle. \tag{2.79}$$

Plugging these into the above we find

$$A_4 = \frac{\langle 4|p_3|1]\langle\hat{1}|p_3|\hat{4}]}{\langle 14\rangle^2[(p_1 + p_2)^2 - m^2]}. \tag{2.80}$$

The numerator may be simplified with a trace identity to

$$\langle 4|p_3|1]\langle\hat{1}|p_3|\hat{4}] = \frac{1}{2}\operatorname{Tr}(\hat{p}_4\hat{p}_3\hat{p}_1\hat{p}_3)$$
$$= 2\left(2(p_3 \cdot \hat{p}_4)(p_3 \cdot \hat{p}_1) - p_3^2(\hat{p}_1 \cdot \hat{p}_4)\right). \tag{2.81}$$

In fact $p_3 \cdot \hat{p}_4 = 0$, which follows from momentum conservation

$$\hat{P}^2 = (p_3 + \hat{p}_4)^2 \quad \Rightarrow \quad m^2 = 2p_3 \cdot \hat{p}_4 + p_3^2 \quad \Rightarrow \quad p_3 \cdot \hat{p}_4 = 0. \tag{2.82}$$

Putting everything together we arrive at the compact expression for the four-point
amplitude

$$A_4\left(1^+, 2_\phi, 3_{\bar\phi}, 4^+\right) = \frac{m^2[14]}{\langle 14\rangle[(p_1 + p_2)^2 - m^2]}, \tag{2.83}$$

which indeed vanishes in the massless limit as promised.

Exercise 2.6 (Four Point Scalar-Gluon Scattering)
Find the four-point massive-scalar-gluon amplitude

$$A_4\left(1^+, 2_\phi, 3_{\bar\phi}, 4^-\right),$$

with one positive and one negative gluon using the above recursive techniques.

2.6 Poincaré and Conformal Symmetry

We now turn to a more conceptual yet important subject: the question how the global symmetries of a gauge field theory manifest themselves at the level of scattering amplitudes. This has proven to be a very rich subject in particular at tree-level. The symmetries of the scattering amplitudes may be grouped into obvious and less obvious symmetries.

The obvious symmetries are the Poincaré transformations under which scattering amplitudes should be invariant. As we are working in the spinor helicity formulation of momentum space the momentum generator $p^{\alpha\dot\alpha}$ is represented by a multiplicative operator

$$p^{\alpha\dot\alpha} = \sum_{i=1}^{n} \lambda_i^\alpha \tilde\lambda_i^{\dot\alpha}, \tag{2.84}$$

and the amplitude $\mathscr{A}_n$ should obey

$$p^{\alpha\dot\alpha} \mathscr{A}_n(\lambda_i, \tilde\lambda_l) = 0. \tag{2.85}$$

This is in fact true in the distributional sense of

$$p\delta(p) = 0, \tag{2.86}$$

thanks to the momentum conserving delta-function present in each amplitude

$$\mathscr{A}_n(\lambda_i, \tilde\lambda_i) = \delta^{(4)}\left(\sum_i p_i\right) A_n(\lambda_i, \tilde\lambda_i). \tag{2.87}$$

The Lorentz generators in the helicity spinor basis come in two pairs of symmetric rank-two tensors $m_{\alpha\beta}$ and $\overline{m}_{\dot\alpha\dot\beta}$ originating from the projections $M^{\mu\nu}(\sigma_{\mu\nu})_{\alpha\beta} = m_{\alpha\beta}$ and $M^{\mu\nu}(\bar\sigma_{\mu\nu})_{\dot\alpha\dot\beta} = \overline{m}_{\dot\alpha\dot\beta}$, see Appendix B for conventions. They are first order differential operators in helicity spinor space,

$$m_{\alpha\beta} = \sum_{i=1}^{n} \lambda_{i(\alpha}\partial_{i\beta)}, \qquad \overline{m}_{\dot\alpha\dot\beta} = \sum_{i=1}^{n} \tilde\lambda_{i(\dot\alpha}\partial_{i\dot\beta)}, \tag{2.88}$$

where $\partial_{i\alpha} := \frac{\partial}{\partial \lambda_i^\alpha}$, $\partial_{i\dot\alpha} := \frac{\partial}{\partial \tilde\lambda_i^{\dot\alpha}}$ and $r_{(\alpha\beta)} := \frac{1}{2}(r_{\alpha\beta} + r_{\beta\alpha})$ denotes symmetrization with unit weight. The invariance of $\mathscr{A}_n(\lambda_i, \tilde\lambda_i)$ under Lorentz-transformations

$$m_{\alpha\beta}\mathscr{A}_n(\lambda_i, \tilde\lambda_i) = 0 = \overline{m}_{\dot\alpha\dot\beta}\mathscr{A}_n(\lambda_i, \tilde\lambda_i) \tag{2.89}$$

is manifest, as it is an immediate consequence of the proper contraction of all Weyl indices within $\mathscr{A}_n$, i.e. the fact that the spinor brackets $\langle ij \rangle$ and $[ij]$ are invariant under $m_{\alpha\beta}$ and $\overline{m}_{\dot\alpha\dot\beta}$. For example,

$$m_{\alpha\beta}\langle jk \rangle = \sum_{i=1}^{n} \lambda_{i(\alpha}\partial_{i\beta)}\lambda_j^\gamma \lambda_{k\gamma} = \lambda_{j\alpha}\lambda_{k\beta} - \lambda_{j\beta}\lambda_{k\alpha} + (\alpha \leftrightarrow \beta) = 0. \tag{2.90}$$

Let us now discuss the less obvious symmetries of $\mathscr{A}_n(\lambda_i, \tilde\lambda_i)$. Classical Yang-Mills theory is invariant under a larger symmetry group than the four-dimensional Poincaré group: Due to the absence of dimensionful parameters in the theory (the coupling g is dimensionless) pure Yang-Mills theory or massless QCD is invariant under a scale transformation

$$x^\mu \to \kappa^{-1}x^\mu, \quad \text{respectively} \quad p^\mu \to \kappa p^\mu. \tag{2.91}$$

The scale transformations of the momenta are generated by the dilatation operator d, whose representation in spinor helicity variables acting on amplitudes is

$$d = \sum_{i=1}^{n} \left(\frac{1}{2}\lambda_i^\alpha \partial_{i\alpha} + \frac{1}{2}\tilde\lambda_i^{\dot\alpha}\partial_{i\dot\alpha} + d_0 \right), \quad d_0 \in \mathbb{R}, \tag{2.92}$$

reflecting the dilatation weight $1/2$ of the λ_i and $\tilde\lambda_i$, i.e. $[d, \lambda_i] = \frac{1}{2}\lambda_i$ and $[d, \tilde\lambda_i] = \frac{1}{2}\tilde\lambda_i$. The constant d_0 is undetermined at this point. It may be fixed by requiring invariance of the MHV amplitudes

$$\mathscr{A}_n^{\text{MHV}} = \delta^{(4)}\left(\sum_i p_i \right) \frac{\langle \lambda_s \lambda_t \rangle^4}{\langle 12 \rangle \cdots \langle n1 \rangle}. \tag{2.93}$$

The dilatation operator d of Eq. (2.92) simply measures the weight in units of mass of the amplitude it acts on plus nd_0

$$d\mathscr{A}_n = \left([\mathscr{A}_n] + nd_0 \right)\mathscr{A}_n. \tag{2.94}$$

We note the weights $[\delta^{(4)}(p)] = -4$, $[\langle \lambda_s \lambda_t \rangle^4] = 4$ and $[\frac{1}{\langle 12 \rangle \cdots \langle n1 \rangle}] = -n$, hence

$$d\mathscr{A}_n^{\text{MHV}} = (-4 + 4 - n + nd_0)\mathscr{A}_n^{\text{MHV}}, \tag{2.95}$$

which vanishes for the choice $d_0 = 1$. One easily checks the invariance under dilatations of the $q\bar{q}gg$-amplitude of Eq. (1.132) and of the $\overline{\text{MHV}}_n$ amplitudes as well.

There is a further symmetry of scale invariant theories, namely the special conformal transformations $k_{\alpha\dot\alpha}$. This is a less obvious symmetry generator realized in terms of a second order differential operator in the spinor variables,

$$k_{\alpha\dot\alpha} = \sum_{i=1}^{n} \partial_{i\alpha} \partial_{i\dot\alpha}. \tag{2.96}$$

Together with the Poincaré and dilatation generators the set $\{p_{\alpha\dot\alpha}, k_{\alpha\dot\alpha}, m_{\alpha\beta}, \overline{m}_{\dot\alpha\dot\beta}, d\}$ generate the conformal group in four dimensions, $SO(2, 4)$.

Let us now prove the invariance of the MHV amplitudes under special conformal transformations. As the only dependence of $\mathscr{A}_n^{\text{MHV}}$ on the conjugate spinors $\tilde\lambda_i$ resides in the momentum conserving delta-function we have

$$k_{\alpha\dot\alpha}\mathscr{A}_n^{\text{MHV}} = \sum_{i=1}^{n} \frac{\partial}{\partial\lambda_i^\alpha} \frac{\partial}{\partial\tilde\lambda_i^{\dot\alpha}} \left(\delta^{(4)}(p) A_n^{\text{MHV}} \right)$$

$$= \sum_{i=1}^{n} \frac{\partial}{\partial\lambda_i^\alpha} \left(\frac{\partial p^{\beta\dot\beta}}{\partial\tilde\lambda_i^{\dot\alpha}} \left(\frac{\partial}{\partial p^{\beta\dot\beta}} \delta^{(4)}(p) \right) A_n^{\text{MHV}} \right)$$

$$= \left[\left(n\frac{\partial}{\partial p^{\alpha\dot\alpha}} + p^{\beta\dot\beta} \frac{\partial}{\partial p^{\beta\dot\alpha}} \frac{\partial}{\partial p^{\alpha\dot\beta}} \right) \delta^{(4)}(p) A_n^{\text{MHV}} \right]$$

$$+ \left(\frac{\partial\delta^{(4)}(p)}{\partial p^{\beta\dot\alpha}} \right) \sum_{i=1}^{n} \lambda_i^\beta \frac{\partial}{\partial\lambda_i^\alpha} A_n^{\text{MHV}}. \tag{2.97}$$

The last term may be rewritten as follows. First, we note the relation

$$\sum_{i=1}^{n} \lambda_{i\alpha} \partial_{i\beta} = \sum_{i=1}^{n} \lambda_{i(\alpha} \partial_{i\beta)} + \frac{1}{2}\varepsilon_{\alpha\beta} \sum_{i} \lambda_i^\gamma \partial_{i\gamma}, \tag{2.98}$$

which follows from decomposing the l.h.s. in a symmetric and anti-symmetric piece. The first term on the right-hand-side is the Lorentz generator $m_{\alpha\beta}$ which we already know annihilates A_n^{MHV}. The remaining term yields

$$\sum_{i=1}^{n} \lambda_i^\beta \frac{\partial}{\partial\lambda_i^\alpha} A_n^{\text{MHV}} = \frac{1}{2}\delta_\alpha^\beta \sum_{i} \lambda_i^\delta \partial_{i\delta} A_n^{\text{MHV}} = (4 - n) A_n^{\text{MHV}}. \tag{2.99}$$

Hence Eq. (2.97) turns into

$$k_{\alpha\dot\alpha}\mathscr{A}_n^{\text{MHV}} = \left[\left(4\frac{\partial}{\partial p^{\alpha\dot\alpha}} + p^{\beta\dot\beta} \frac{\partial}{\partial p^{\beta\dot\alpha}} \frac{\partial}{\partial p^{\alpha\dot\beta}} \right) \delta^{(4)}(p) \right] A_n^{\text{MHV}}. \tag{2.100}$$

Indeed in a distributional sense we have

$$p^{\beta\dot\beta} \frac{\partial}{\partial p^{\beta\dot\alpha}} \frac{\partial}{\partial p^{\alpha\dot\beta}} \delta^{(4)}(p) = -4\frac{\partial}{\partial p^{\alpha\dot\alpha}} \delta^{(4)}(p), \tag{2.101}$$

which one sees by integrating the second derivative expression against a test function $F(p)$,

$$\int d^4 p\, F(p)\, p^{\beta\dot\beta}\, \frac{\partial}{\partial p^{\beta\dot\alpha}}\, \frac{\partial}{\partial p^{\alpha\dot\beta}}\, \delta^{(4)}(p)$$

$$= \int d^4 p\left(\left[\frac{\partial}{\partial p^{\beta\dot\alpha}} F(p)\right] 2\delta_\alpha^\beta + \left[\frac{\partial}{\partial p^{\alpha\dot\beta}} F(p)\right] 2\delta_{\dot\alpha}^{\dot\beta}\right)$$

$$= 4\int d^4 p\left[\frac{\partial}{\partial p^{\alpha\dot\alpha}} F(p)\right]\delta^{(4)}(p). \tag{2.102}$$

This proves the vanishing of $k_{\alpha\dot\alpha}\mathscr{A}_n^{\mathrm{MHV}}$, as claimed.

Summarizing, we have constructed a representation of the conformal group whose generators are represented by differential operators of degree one ($m_{\alpha\beta}$, $\overline{m}_{\dot\alpha\dot\beta}$, d), of degree two ($k_{\alpha\dot\alpha}$) and as a multiplicative operator ($p_{\alpha\dot\alpha}$) in an n-particle helicity spinor space. This representation is natural as amplitudes are functions in this space. All the generators leave the scattering amplitudes invariant. We have verified this explicitly for the MHV amplitudes. The representation obeys the commutation relations of the conformal algebra $\mathfrak{so}(2,4)$

$$\left[d, p^{\alpha\dot\alpha}\right] = p^{\alpha\dot\alpha}, \qquad [d, k_{\alpha\dot\alpha}] = -k_{\alpha\dot\alpha}, \qquad [d, m_{\alpha\beta}] = 0 = [d, \overline{m}_{\dot\alpha\dot\beta}],$$

$$\left[k_{\alpha\dot\alpha}, p^{\beta\dot\beta}\right] = \delta_\alpha^\beta \delta_{\dot\alpha}^{\dot\beta} d + m_\alpha{}^\beta \delta_{\dot\alpha}^{\dot\beta} + \overline{m}_{\dot\alpha}{}^{\dot\beta}\delta_\alpha^\beta, \tag{2.103}$$

plus the Poincaré commutators discussed in Sect. 1.1.

The origin of this helicity spinor space representation becomes clear if one looks at the more familiar representation of the conformal group in configuration space x^μ which reads ($\partial_\mu := \frac{\partial}{\partial x^\mu}$)

$$M_{\mu\nu} = i(x_\mu \partial_\nu - x_\nu \partial_\mu), \qquad P_\mu = -i\partial_\mu,$$

$$D = -ix^\mu \partial_\mu, \qquad K_\mu = i\left(x^2 \partial_\mu - 2x_\mu x^\nu \partial_\nu\right). \tag{2.104}$$

A Fourier transform $\int d^4 x\, e^{ip\cdot x}\, \mathscr{O}(x, \partial_x)$ brings this representation into momentum space, which in turn can be mapped to the helicity spinor representation discussed in Sect. 1.6. From this point of view it is clear why $p^{\alpha\dot\alpha}$ becomes a multiplication operator and $k_{\alpha\dot\alpha}$ a second order derivative operator in momentum space.

Summary: Conformal Generators   Here we collect the generators of the conformal algebra. For simplicity of notation, we write their single-particle action.

$$p^{\alpha\dot\alpha} = \lambda^\alpha \tilde\lambda^{\dot\alpha}, \qquad k_{\alpha\dot\alpha} = \partial_\alpha \partial_{\dot\alpha},$$

$$m_{\alpha\beta} = \lambda_{(\alpha}\partial_{\beta)} := \frac{1}{2}(\lambda_\alpha \partial_\beta + \lambda_\beta \partial_\alpha), \qquad \overline{m}_{\dot\alpha\dot\beta} = \tilde\lambda_{(\dot\alpha}\partial_{\dot\beta)}, \tag{2.105}$$

$$d = \frac{1}{2}\lambda^\alpha \partial_\alpha + \frac{1}{2}\tilde\lambda^{\dot\alpha}\partial_{\dot\alpha} + 1.$$

The helicity generator is given by $h = -\frac{1}{2}\lambda^\alpha \partial_\alpha + \frac{1}{2}\tilde{\lambda}^{\dot\alpha}\partial_{\dot\alpha}$. It commutes with all generators of the conformal algebra.

Exercise 2.7 (Conformal Algebra)

Show that the representation constructed in the above Eq. (2.105) indeed obeys the commutation relations of the conformal algebra given in Eq. (2.103).

Exercise 2.8 (Inversion and Special Conformal Transformations)

The generator K_μ of Eq. (2.104) generates infinitesimal special conformal transformations. A finite special conformal transformation is given by

$$K^\mu: \quad x^\mu \to x'^\mu = \frac{x^\mu - a^\mu x^2}{1 - 2a \cdot x + a^2 x^2}, \quad a^\mu : \text{transformation parameter.} \quad (2.106)$$

An intuition on the character of these transformations may be found by noting that the action of K^μ may be also written as $K^\mu = I P^\mu I$, i.e. an inversion $I x^\mu = \frac{x^\mu}{x^2}$ followed by a translation by a^μ followed by another inversion. Show that $K^\mu = I P^\mu I$ is equivalent to Eq. (2.106).

2.7 $\mathcal{N} = 4$ Super Yang-Mills Theory

So far we have mostly discussed pure Yang-Mills theory or massless QCD. Our external states were either gluons ($h = \pm 1$) or quarks ($h = \pm 1/2$). However, a renormalizable quantum field theory in four dimensions could also contain scalar fields with helicity $h = 0$. Of course the Higgs is an example for an elementary scalar particle realized in Nature. In particular if we continue to insist on conformal symmetry at tree-level, i.e. the absence of any dimensionful quantity in the bare Lagrangian of the theory, then the only allowed terms for a massless scalar in the Lagrangian are quartic.

Interestingly, irrespective of the details of the content and couplings of the fermionic and scalar matter fields in a gauge theory, the n-gluon *tree* level amplitudes are identical to the pure Yang-Mills theory ones. This is the case because scalars and fermions interact with each other and the gauge field via vertices of the type

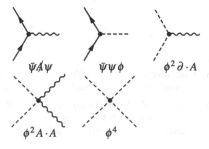

Fig. 2.8 Tree-level and one-loop examples for a gauge field theory coupled to scalars and fermions: at tree-level a scalar of fermion line always has to exit the diagram

Hence in a tree-level diagram with only external gluon legs, scalars or fermions cannot appear: they are always produced in pairs from gluon lines and thus have to exit the diagram at tree-level, see Fig. 2.8. Thus as long as one is interested in pure gluon amplitudes at tree-level[1] one may assemble the matter content of the gauge theory according to one's preference. It then proves useful to maximize the amount of symmetries of the theory in this choice. There is a distinguished and very special gauge theory which surpasses all others in its remarkable properties: it is the *maximally supersymmetric* Yang-Mills theory or $\mathcal{N} = 4$ super Yang-Mills theory, which we now introduce. It may be thought of as a supersymmetric version of QCD.

Supersymmetry is a very attractive concept in elementary particle physics albeit not yet discovered in Nature. It proposes a symmetry between fermionic and bosonic fields generated by Grassmann odd supersymmetry generators leading to a graded (supersymmetric) extension of the Poincaré algebra. In its simplest ($\mathcal{N} = 1$) gauge theoretical version there is one super-partner to every gluon A_μ^a: the spin $1/2$ gluino ψ_α^a. Indeed, the supersymmetry of gauge field theories can be extended to a maximal degree of four ($\mathcal{N} = 4$ supersymmetry):[2] one then has *four* gluinos $\psi_{\alpha A}^a$ ($A = 1, 2, 3, 4$) as the super-partners to the gluon field A_μ^a. Closure of the supersymmetry algebra requires the additional presence of six real scalar fields $\phi^{aAB} = -\phi^{aBA}$. In total the field content of the $\mathcal{N} = 4$ super Yang-Mills theory (we consider again the $SU(N)$ gauge group) is

$$A_\mu^a: \qquad \text{gluon } a = 1, \dots, N^2 - 1,$$

$$\psi_{\alpha A}^a, \bar{\psi}^{\dot{\alpha}aA}: \qquad \text{4 gluinos } \alpha, \dot{\alpha} = 1, 2; \; A = 1, 2, 3, 4,$$

$$\phi^{aAB}: \qquad \text{6 scalars antisymmetric in } AB.$$

For the scalars and gluinos we have the complex conjugation properties

$$\left(\phi^{AB}\right)^* = \phi_{AB} = \frac{1}{2}\varepsilon_{ABCD}\phi^{CD}, \qquad \left(\psi_{\alpha A}^a\right)^* = \bar{\psi}_{\dot{\alpha}}^{aA}. \qquad (2.107)$$

---

[1]One can also show that QCD tree-level amplitudes with quarks and gluons can be deduced from their supersymmetric cousins, at least for up to four quark-antiquark pairs.

[2]In fact one may extend beyond $\mathcal{N} = 4$ supersymmetries at the prize of leaving the realm of renormalizable quantum field theories. This leads to supergravity with a graviton and gravitinos in the spectrum. The maximally extended supersymmetric model is then $\mathcal{N} = 8$ supergravity.

The extended supersymmetry also forces all fields to transform in the adjoint representation, distinguishing the gluinos from the quarks, which transform in the fundamental representation. Nevertheless, this difference is not visible at the level of color-ordered amplitudes where all gauge group dependences have been stripped off. The $\mathcal{N} = 4$ super Yang-Mills action reads

$$S = \frac{1}{g_{YM}^2} \int d^4x \, \mathrm{Tr}\left(-\frac{1}{4}F_{\mu\nu}^2 - (D_\mu \phi_{AB})D^\mu \phi^{AB} - \frac{1}{2}[\phi_{AB}, \phi_{CD}][\phi^{AB}, \phi^{CD}] \right.$$

$$\left. + i\bar{\psi}_{\dot{\alpha}}^A \sigma_\mu^{\dot{\alpha}\alpha} D^\mu \psi_{\alpha A} - \frac{i}{2}\psi_A^\alpha [\phi^{AB}, \psi_{\alpha B}] - \frac{i}{2}\bar{\psi}_{\dot{\alpha}}^A [\phi_{AB}, \bar{\psi}^{\dot{\alpha}B}] \right). \tag{2.108}$$

Its form is uniquely fixed by the $\mathcal{N} = 4$ supersymmetry. For a review and more details, see Ref. [8]. There are only two tunable parameters: the gauge coupling g_{YM} and the rank N of the gauge group $SU(N)$. Notably in $\mathcal{N} = 4$ super Yang-Mills the gauge coupling g_{YM} is not renormalized at the quantum level, i.e. the theory is ultraviolet finite. In other words, the conformal symmetry at tree-level survives the quantization process without anomalies, and we have an interacting four dimensional quantum conformal field theory.

In fact this does not imply that there are no divergences arising in scattering amplitudes and correlation functions in this theory: radiative corrections to scattering amplitudes suffer from infrared (IR) divergences, just as in QCD, but are free of ultraviolet (UV) divergences reflecting the non-renormalization of the coupling constant.[3] Also composite gauge invariant operators, such as $\mathrm{Tr}(\phi_{AB}\phi^{AB})$, are renormalized in order to cure their inherent short-distance UV divergences. This induces anomalous scaling dimensions which are non-trivial functions of g_{YM}.

It may be said that $\mathcal{N} = 4$ super Yang-Mills is the interacting four dimensional quantum field theory with the largest amount of symmetry: maximally extended supersymmetry, quantum conformal symmetry and local $SU(N)$ gauge invariance.

We shall be interested in tree-level and loop-level color-ordered amplitudes in this highly symmetric quantum field theory, which we shall analyze in the following.

2.7.1 On-shell Superspace and Superfields

In a supersymmetric theory the on-shell degrees of freedom are balanced between bosons and fermions. In the $\mathcal{N} = 4$ super Yang-Mills (SYM) model we have 8 bosonic and 8 fermionic on-shell degrees of freedom, which may be grouped in Table 2.1.

---

[3]Depending on the formalism used, the elementary fields may need to be renormalized.

Table 2.1 The $\mathcal{N} = 4$ super Yang-Mills on-shell field content

Field	Bosons			Fermions	
	g_+	g_-	S_{AB}	$\tilde{g}_A$	$\bar{\tilde{g}}^A$
Name	gluon		scalar	gluino	anti-gluino
Helicity	$+1$	-1	0	$+1/2$	$-1/2$
Degrees of freedom	1	1	6	4	4
$SU(4)_R$ representation	singlet		anti-symmetric (**6**)	fundamental (**4**)	anti-fund ($\bar{\mathbf{4}}$)

This $\mathcal{N} = 4$ SYM on-shell multiplet may be assembled into one on-shell superfield Φ upon introducing the Grassmann odd parameter η^A with $A = 1, 2, 3, 4$

$$\Phi(p, \eta) = g_+(p) + \eta^A \tilde{g}_A(p) + \frac{1}{2!}\eta^A \eta^B S_{AB}(p)$$

$$+ \frac{1}{3!}\eta^A \eta^B \eta^C \varepsilon_{ABCD}\bar{\tilde{g}}^D(p) + \frac{1}{4!}\eta^A \eta^B \eta^C \eta^D \varepsilon_{ABCD} g_-(p). \quad (2.109)$$

If we assign the helicity $h = 1/2$ to the Grassmann variable η^A then the on-shell superfield $\Phi(\eta)$ carries uniform helicity $h = 1$. This extends our definition Eq. (1.75) for the helicity operator h to the supersymmetric case

$$h = \frac{1}{2}\big[-\lambda^\alpha \partial_\alpha + \tilde{\lambda}^{\dot\alpha} \partial_{\dot\alpha} + \eta^A \partial_A\big], \qquad \partial_A := \frac{\partial}{\partial \eta^A}, \quad (2.110)$$

with $h\Phi(\eta) = \Phi(\eta)$. The introduced superspace $\{\lambda^\alpha, \tilde{\lambda}^{\dot\alpha}, \eta^A\}$ is chiral in the following sense: the complex conjugate of η^A is not part of the superspace: $\overline{(\eta^A)} = \bar{\eta}_A$.

It is natural to consider color ordered superamplitudes in $\mathcal{N} = 4$ SYM whose external legs are parametrized by a point in super-momentum space $\Lambda_i := \{\lambda_i, \tilde{\lambda}_i, \eta_i\}$ associated to an on-shell superfield $\Phi(\Lambda_i)$, i.e.

$$\mathbb{A}_n\big(\{\lambda_i, \tilde{\lambda}_i, \eta_i\}\big) = \big\langle \Phi(\lambda_1, \tilde{\lambda}_1, \eta_1) \cdots \Phi(\lambda_n, \tilde{\lambda}_n, \eta_n)\big\rangle. \quad (2.111)$$

This prescription packages all possible component field amplitudes involving gluons, gluinos and scalars as external states into a single object. The component level amplitudes may then be extracted from a known $\mathbb{A}_n(\{\lambda_i, \tilde{\lambda}_i, \eta_i\})$ upon expanding it in the Grassmann odd η_i^A variables.

For example, the expansion of the η_i^A-polynomial of $\mathbb{A}_n(\Lambda_i)$ will contain terms such as

$$
\begin{aligned}
&(\eta_1)^4 (\eta_2)^4 \mathscr{A}_n(-, -, +, \ldots, +) \quad \text{with } \eta_i^4 := \frac{1}{4!}\varepsilon_{ABCD}\eta_i^A \eta_i^B \eta_i^C \eta_i^D, \\
&(\eta_1^4)\varepsilon_{ACDE}\eta_2^C \eta_2^D \eta_2^E \eta_3^B \mathscr{A}_n\left(-, \tilde{g}^A, \bar{\tilde{g}}_B, +, \ldots, +\right).
\end{aligned}
\tag{2.112}
$$

Here $\mathscr{A}_n$ denotes the resulting component field amplitudes, in this example two MHV amplitudes, namely a pure gluon and a gluon-$\tilde{g}\bar{\tilde{g}}$ amplitude. We also stress the property

$$
\begin{aligned}
&h_i \mathbb{A}_n(1, \ldots, n) = \mathbb{A}_n(1, \ldots, n), \\
&h_i = \frac{1}{2}\left[-\lambda_i^\alpha \partial_{i\alpha} + \tilde{\lambda}_i^{\dot\alpha} \partial_{i\dot\alpha} + \eta_i^A \partial_{iA}\right], \quad \forall i = 1, \ldots, n,
\end{aligned}
\tag{2.113}
$$

which holds 'locally' for each individual leg. This is a consequence of the uniform helicity 1 of any on-shell superfield $\Phi(\eta)$.

2.7.2 Superconformal Symmetry

The $\mathcal{N} = 4$ supersymmetry transformations are generated by the operators $q^{\alpha A}$ and $\bar{q}_A^{\dot\alpha}$. By definition their anti-commutator yields the translation operator

$$
\{q^{\alpha A}, \bar{q}_B^{\dot\alpha}\} = \delta_B^A p^{\alpha\dot\alpha},
\tag{2.114}
$$

which is the key commutation relation of supersymmetry: The supersymmetry transformation may be thought of as the 'square-root' of the translation. As $p^{\alpha\dot\alpha} = \lambda^\alpha \tilde{\lambda}^{\dot\alpha}$ the natural representation of the supersymmetry generators in our on-shell superspace then is

$$
q^{\alpha A} = \lambda^\alpha \eta^A, \qquad \bar{q}_A^{\dot\alpha} = \tilde{\lambda}^{\dot\alpha} \frac{\partial}{\partial \eta^A},
\tag{2.115}
$$

manifestly obeying Eq. (2.114). This representation enables us to read off the supersymmetry q-transformations of the on-shell components of the superfield $\Phi(p, \eta)$

$$
\delta_q \Phi(p, \eta) = \xi_{\alpha A}\left(q^{\alpha A} \Phi(p, \eta)\right),
\tag{2.116}
$$

with the Grassmann odd transformation parameter $\xi_{\alpha A}$. The respective left- and right-hand sides of this equation then take the form

$$\delta_q \Phi(p, \eta)$$

$$= \delta_q g_+ + \eta^A \delta_q \tilde{g}_A(p) + \frac{1}{2!} \eta^A \eta^B \delta_q S_{AB}(p)$$

$$+ \frac{1}{3!} \eta^A \eta^B \eta^C \varepsilon_{ABCD} \delta_q \bar{\tilde{g}}^D(p) + \frac{1}{4!} \eta^A \eta^B \eta^C \eta^D \varepsilon_{ABCD} \delta_q g_-(p),$$

$$\xi_{\alpha A}\left(q^{\alpha A} \Phi(p, \eta)\right)$$

$$= \xi_{\alpha A} \lambda^\alpha \left(\eta^A g_+ + \eta^A \eta^B \tilde{g}_B + \frac{1}{2!} \eta^A \eta^B \eta^C S_{BC} \right.$$

$$\left. + \frac{1}{3!} \eta^A \eta^B \eta^C \eta^D \varepsilon_{BCDE} \bar{\tilde{g}}^E \right).$$

$$(2.117)$$

This result implies the q-variations of the component on-shell fields by comparing the left-hand and right-hand side of Eq. (2.116)

$$\delta_q g_+ = 0, \qquad \delta_q \tilde{g}_A = -\langle \xi_A \lambda \rangle g_+, \qquad \delta_q S_{AB} = -\langle \xi_A \lambda \rangle \tilde{g}_B + \langle \xi_B \lambda \rangle \tilde{g}_A,$$

$$\delta_q \bar{\tilde{g}}^A = \varepsilon^{ABCD} \langle \xi_B \lambda \rangle S_{CD}, \qquad \delta_q g_- = -\langle \xi_A \lambda \rangle \bar{\tilde{g}}^A.$$

$$(2.118)$$

With the same method one establishes the $\bar{q}$-variations

$$\delta_{\bar{q}} g_+ = [\tilde{\lambda} \tilde{\xi}^A] \tilde{g}_A, \qquad \delta_{\bar{q}} \tilde{g}_A = [\tilde{\lambda} \tilde{\xi}^B] S_{BA},$$

$$\delta_{\bar{q}} S_{AB} = \varepsilon_{ABCD} [\tilde{\lambda} \tilde{\xi}^C] \bar{\tilde{g}}^D,$$

$$\delta_{\bar{q}} \bar{\tilde{g}}^A = -[\tilde{\lambda} \tilde{\xi}^A] g_-, \qquad \delta_{\bar{q}} g_- = 0,$$

$$(2.119)$$

with the Grassmann odd parameter $\tilde{\xi}_{\dot{\alpha}}^A$.

Having established the supersymmetry generators, let us now discuss the remaining generators of the superconformal symmetry algebra. In addition to the known Lorentz symmetry generators $m_{\alpha\beta}$ and $\overline{m}_{\dot{\alpha}\dot{\beta}}$, discussed in Sect. 2.6, one now has an additional global $SU(4)$ R-symmetry generated by $r^A{}_B$ which acts as an internal rotation in the η^A space

$$r^A{}_B = \eta^A \partial_B - \frac{1}{4} \delta^A_B \eta^C \partial_C, \qquad \partial_A := \frac{\partial}{\partial \eta^A},$$

$$m_{\alpha\beta} = \lambda_{(\alpha} \partial_{\beta)}, \qquad \overline{m}_{\dot{\alpha}\dot{\beta}} = \tilde{\lambda}_{(\dot{\alpha}} \tilde{\partial}_{\dot{\beta})}.$$

$$(2.120)$$

In the conformal sector the generator of special conformal transformations of Sect. 2.6

$$k_{\alpha\dot{\alpha}} = \partial_\alpha \tilde{\partial}_{\dot{\alpha}}$$

$$(2.121)$$

is augmented by two superconformal symmetry generators $s_{\alpha A}$ and $\bar{s}^A_{\dot\alpha}$ which arise from the commutators

$$[k_{\alpha\dot\alpha}, q^{\beta A}] = \delta^\beta_\alpha \bar{s}^A_{\dot\alpha}, \qquad \bar{s}^A_{\dot\alpha} = \eta^A \partial_{\dot\alpha},$$

$$[k_{\alpha\dot\alpha}, \bar{q}^{\dot\beta}_A] = \delta^{\dot\beta}_{\dot\alpha} s_{\alpha A}, \qquad s_{\alpha A} = \partial_\alpha \partial_A. \tag{2.122}$$

The complete super-conformal symmetry algebra reads

$$\{q^{\alpha A}, \bar{q}^{\dot\alpha}_B\} = \delta^A_B p^{\alpha\dot\alpha}, \qquad \{s_{\alpha A}, \bar{s}^B_{\dot\alpha}\} = \delta^B_A k_{\alpha\dot\alpha},$$

$$\{q^{\alpha A}, s_{\beta B}\} = m^\alpha{}_\beta \delta^A_B + \delta^\alpha_\beta r^A{}_B + \frac{1}{2}\delta^\alpha_\beta \delta^A_B (d + c),$$

$$\{\bar{q}^{\dot\alpha}_A, \bar{s}^B_{\dot\beta}\} = \overline{m}^{\dot\alpha}{}_{\dot\beta} \delta^A_B - \delta^{\dot\alpha}_{\dot\beta} r^B{}_A + \frac{1}{2}\delta^{\dot\alpha}_{\dot\beta} \delta^B_A (d - c), \tag{2.123}$$

$$[p^{\alpha\dot\alpha}, s_{\beta A}] = \delta^\alpha_\beta \bar{q}^{\dot\alpha}_A, \qquad [p^{\alpha\dot\alpha}, \bar{s}^A_{\dot\beta}] = \delta^{\dot\alpha}_{\dot\beta} q^{\alpha A},$$

with the dilatation generator and central charge

$$d = \frac{1}{2}[\lambda^\alpha \partial_\alpha + \tilde\lambda^{\dot\alpha} \partial_{\dot\alpha} + 1], \qquad c = 1 + \frac{1}{2}(\lambda^\alpha \partial_\alpha - \tilde\lambda^{\dot\alpha} \partial_{\dot\alpha} - \eta^A \partial_A) = 1 - h. \tag{2.124}$$

While c commutes with all generators of the above algebra, d measures their weight in momentum units. Also note that the central charge vanishes on super-amplitudes $c_i \mathbb{A}_n(\Lambda_i) = 0$ locally for every leg as a consequence of Eq. (2.113). Together with the conformal algebra

$$[k_{\alpha\dot\alpha}, p^{\beta\dot\beta}] = \delta^\beta_\alpha \delta^{\dot\beta}_{\dot\alpha} d + m_\alpha{}^\beta \partial^{\dot\beta}_{\dot\alpha} + \overline{m}_{\dot\alpha}{}^{\dot\beta} \delta^\beta_\alpha,$$

$$[d, p^{\alpha\dot\alpha}] = p^{\alpha\dot\alpha}, \qquad [d, k_{\alpha\dot\alpha}] = -k_{\alpha\dot\alpha}, \qquad [c, *] = 0, \tag{2.125}$$

and the Poincaré algebra this constitutes the super-conformal $\mathfrak{psu}(2, 2|4)$ symmetry algebra. Super-amplitudes are invariant under all these generators.

2.7.3 Super-amplitudes, and Extraction of Components

The super-amplitudes are invariant under the superconformal symmetry algebra $\mathfrak{psu}(2, 2|4)$ discussed above. As before the symmetry generators acting on n-leg super-amplitudes are given by the sum of the single-particle representations

$$p^{\alpha\dot\alpha} = \sum_{i=1}^n \lambda^\alpha_i \tilde\lambda^{\dot\alpha}_i, \qquad q^{\alpha A} = \sum_{i=1}^n \lambda^\alpha_i \eta^A_i, \qquad \bar{q}^{\dot\alpha}_A = \sum_{i=1}^n \tilde\lambda^{\dot\alpha}_i \partial_{iA}, \qquad \text{etc.}, \tag{2.126}$$

and we have the conservation of total charges in the sense of

$$\{p^{\alpha\dot\alpha}, d, k_{\alpha\dot\alpha}, m_{\alpha\beta}, \overline{m}_{\alpha\beta}, r^A{}_B; q^{\alpha A}, \bar{q}^{\dot\alpha}_A, s_{\alpha A}, \bar{s}^B_{\dot\alpha}; h_i\} \circ \mathbb{A}_n = 0. \tag{2.127}$$

Note that only $p^{\alpha\dot\alpha}$ and $q^{\alpha A}$ act multiplicatively, whereas the set of generators $\{d, m_{\alpha\beta}, \overline{m}_{\dot\alpha\dot\beta}, r^A{}_B; \bar{q}^{\dot\alpha}_A, \bar{s}^B_{\dot\alpha}; h_i\}$ are first order differential operators, while $k_{\alpha\dot\alpha}$ and $s_{\alpha A}$ are second order differential operator in the variables $\{\lambda^\alpha_i, \tilde\lambda^{\dot\alpha}_i, \eta^A_i\}$.

The invariance under the multiplicatively represented operators $p^{\alpha\dot\alpha}$ and $q^{\alpha A}$ then requires the general form of the super-amplitudes to be (for $n > 3$)

$$\mathbb{A}_n(\lambda_i, \tilde\lambda_i, \eta_i) = \frac{\delta^{(4)}(p)\delta^{(8)}(q)}{\langle 12\rangle\langle 23\rangle\cdots\langle n1\rangle} P_n(\lambda_i, \tilde\lambda_i, \eta_i), \tag{2.128}$$

enforcing the p and q conservation through delta functions. In the above the analytic function P_n is arbitrary and we took out the MHV-like numerator factor $\langle 12\rangle\cdots\langle n1\rangle$ from the definition of P_n. This is a pure convention.

Due to the $\mathfrak{su}(4)$ R-symmetry $P_n(\lambda_i, \tilde\lambda_i, \eta_i)$ has an η-expansion of the form

$$P_n(\lambda_i, \tilde\lambda_i, \eta_i) = \underset{\text{MHV}}{\overset{\displaystyle P_n^{(0)}}{\updownarrow}} + \underset{\text{NMHV}}{\overset{\displaystyle P_n^{(4)}}{\updownarrow}} + \underset{\text{N}^2\text{MHV}}{\overset{\displaystyle P_n^{(8)}}{\updownarrow}} + \cdots + \underset{\overline{\text{MHV}}}{\overset{\displaystyle P_n^{(4n-16)}}{\updownarrow}}, \tag{2.129}$$

where $P_n^{(l)} \sim \mathcal{O}(\eta^l)$, which corresponds directly to the indicated helicity classification. We shall see shortly that $P_n^{(0)} = 1$.

As we encounter them for the first time, let us briefly discuss the formalism of Grassmann odd delta-functions. For a single real Grassmann odd variable θ we have the integration rules

$$\int d\theta \cdot \theta = 1, \qquad \int d\theta \cdot 1 = 0, \quad \text{thus} \int d\theta \,\hat{=}\, \frac{\partial}{\partial\theta}. \tag{2.130}$$

Indeed $\delta(\theta) = \theta$ as one sees upon integrating against a test-function $F(\theta) = F_0 + \theta F_1$

$$\int d\theta\delta(\theta - \theta_0)F(\theta) = \int d\theta(\theta - \theta_0)(F_0 + \theta F_1) = \int d\theta(-\theta_0 F_0 + \theta(F_0 + \theta_0 F_1)$$

$$= F_0 + \theta_0 F_1 = F(\theta_0). \tag{2.131}$$

We also note the integral representation

$$\delta(\theta - \theta_0) = \int d\bar\theta e^{\bar\theta(\theta - \theta_0)}. \tag{2.132}$$

The eight-dimensional delta function in Eq. (2.128) refers to the explicit expression

$$\delta^{(8)}(q) = \delta^{(8)}\left(\sum_{i=1}^n \lambda^\alpha_i \eta^A_i\right) := \prod_{\alpha=1}^2 \prod_{A=1}^4 \left(\sum_{i=1}^n \lambda^\alpha_i \eta^A_i\right) \sim \mathcal{O}(\eta^8). \tag{2.133}$$

Hence, $\mathbb{A}_n$ has an η-expansion starting at order η^8, which in turn implies that large classes of component-field amplitudes vanish. In particular we see that

$$\mathbb{A}_n|_{\eta^0} = 0 \quad \Rightarrow \quad \mathscr{A}_n^{\text{gluon}}\left(1^+, 2^+, \ldots, n^+\right) = 0$$

$$\text{from 1 term,} \tag{2.134}$$

$$\mathbb{A}_n|_{\eta^4} = 0 \quad \Rightarrow \quad \mathscr{A}_n^{\text{gluon}}\left(1^-, 2^+, \ldots, n^+\right) = 0$$

$$\text{from } \varepsilon_{ABCD} \eta_1^A \eta_1^B \eta_1^C \eta_1^D \text{ term,} \tag{2.135}$$

$$\Rightarrow \quad \mathscr{A}_n^{\phi\phi g^{n-2}}\left(1_\phi, 2_\phi, 3^+, \ldots, n^+\right) = 0$$

$$\text{from } \varepsilon_{ABCD} \eta_1^A \eta_1^B \eta_2^C \eta_2^D \text{ term,} \tag{2.136}$$

$$\Rightarrow \quad \mathscr{A}_n^{\bar{\tilde{g}}\tilde{g}g^{n-2}}\left(1_{\tilde{g}_A}^+, 2_{\bar{\tilde{g}}^A}^-, 3^+, \ldots, n^+\right) = 0$$

$$\text{from } \varepsilon_{ABCD} \eta_1^A \eta_2^B \eta_2^C \eta_2^D \text{ term.} \tag{2.137}$$

Moreover note that due to the $\mathfrak{su}(4)$ R-symmetry of the super-amplitudes the η-expansion of $\mathbb{A}_n$ is an expansion in powers of η^4. These results carry over to the following vanishing tree-amplitudes in massless QCD

$$\mathscr{A}_{g^n}^{\text{QCD}}\left(1^\pm, 2^+, 3^+, \ldots, n^+\right) = 0, \tag{2.138}$$

$$\mathscr{A}_{\bar{q}qg^{n-2}}^{\text{QCD}}\left(1_q^+, 2^+, \ldots, (i-1)^+, i_{\bar{q}}^-, (i+1)^+, \ldots, n^+\right) = 0, \tag{2.139}$$

reproducing the results of Sect. 1.11 based on an explicit evaluation of the color-ordered Feynman graphs. As promised we now understand this vanishing as following from a hidden supersymmetry for QCD tree-amplitudes with external gluon legs and up to one quark-anti-quark line. We also see the vanishing of the scalar-gluon amplitudes when all gluons have positive helicity, as shown for $n = 4$ in Sect. 2.5.

The alert reader may object that there could be internal scalar contributions to the $\mathcal{N} = 4$ tree-amplitudes which would invalidate the equivalence to massless QCD-trees. However, scalars may only be exchanged between gluino lines at tree-level due to the absence of a ϕAA interaction in the theory. Hence, as long as there is only one gluino line running through the tree-diagram with otherwise external gluon legs, there can be no intermediate scalar exchange. In fact, this property generalizes to an arbitrary number of gluino lines, as long as they are all of the same flavor: the $\mathcal{N} = 4$ Yukawa vertices of Eq. (2.108) are of the type $\psi_A \psi_B \phi^{AB}$ or $\bar{\psi}^A \bar{\psi}^{\alpha B} \phi_{AB}$ respectively. These clearly vanish for $A = B$. In this sense massless QCD at tree-level is effectively $\mathcal{N} = 1$ supersymmetric.

The component amplitudes can be extracted in an elegant way by carrying out integrations over the Grassmann parameters. The latter are always localized thanks to corresponding Grassmann delta functions in the amplitudes, such that in general extracting component amplitudes from superamplitudes amounts to linear algebra. We will see examples of this in exercises at the end of this chapter.

2.7.4 Super BCFW-Recursion

We now want to construct a super-amplitude formulation of the BCFW-recursion relations generalizing the results of Sect. 2.1. Such a recursion should exist as the super-amplitudes merely represent a packaging of component field amplitudes with the help of the Grassmann odd η_i's and the component field amplitudes do enjoy a BCFW-recursion relation. For the derivation of the BCFW-recursion in Sect. 2.1 we considered the complex shifts of the helicity spinors at neighboring legs 1 and n

$$
\begin{aligned}
\lambda_1 &\to \lambda_1 - z\lambda_n = \hat{\lambda}_1, \\
\tilde{\lambda}_n &\to \tilde{\lambda}_n + z\tilde{\lambda}_1 = \hat{\tilde{\lambda}}_n.
\end{aligned}
\tag{2.140}
$$

Recall that this shift preserves the total momentum

$$
\begin{aligned}
p_1 &\to \hat{p}_1 = \lambda_1\tilde{\lambda}_1 - z\lambda_n\tilde{\lambda}_1, \quad p_n \to \hat{p}_n = \lambda_n\tilde{\lambda}_n + z\lambda_n\tilde{\lambda}_1, \\
&\Rightarrow \quad \hat{p}_1 + \hat{p}_n = p_1 + p_n.
\end{aligned}
\tag{2.141}
$$

For the super-amplitude it is natural to consider an additional z-dependent shift in the η_i^A variables at leg positions 1 and n. The question is which one to take. Again the guideline is the conservation of 'super-momentum' $q^{\alpha A}$ of the superamplitude $q^{\alpha A} = \sum_{i=1}^n \lambda_i^\alpha \eta_i^A$ whose leg 1 and n components transform under the shifts Eq. (2.140) as

$$
\begin{aligned}
q_1^{\alpha A} &\to \hat{q}_1^{\alpha A} = (\lambda_1 - z\lambda_n)\hat{\eta}_1, \\
q_n^{\alpha A} &\to \hat{q}_n^{\alpha A} = \lambda_n\hat{\eta}_n.
\end{aligned}
\tag{2.142}
$$

with so far unspecified shifted $\hat{\eta}_1$ and $\hat{\eta}_n$. The unique choice to preserve total $q^{\alpha A}$ is

$$
\begin{aligned}
\hat{\eta}_1 &= \eta_1, \quad \hat{\eta}_n = \eta_n + z\eta_1, \\
&\Rightarrow \quad \hat{q}_1^{\alpha A} + \hat{q}_n^{\alpha A} = q_1^{\alpha A} + q_n^{\alpha A}.
\end{aligned}
\tag{2.143}
$$

The derivation of the super-BCFW recursion then follows the same steps as in Sect. 2.1: the unshifted super-amplitude $\mathbb{A}_n(z=0)$ results from the knowledge of the poles of $\mathbb{A}_n(z)$ in the complex plane where the amplitude factorizes into a left $\mathbb{A}^L$ and right $\mathbb{A}^R$ part. The super-recursion relation reads

$$
\mathbb{A}_n(1,\ldots,n) = \sum_{i=3}^{n-1} \int d^4\eta_{\hat{P}_i}\, \mathbb{A}_i^L\big(\hat{1}(z_{P_i}), 2, \ldots, i-1, -\hat{P}(z_{P_i})\big)
$$

$$
\times \frac{1}{P_i^2}\mathbb{A}_{n-i+2}^R\big(\hat{P}(z_{P_i}), i, \ldots, n-1, \hat{n}(z_{P_i})\big) + \mathrm{Res}(z=\infty),
\tag{2.144}
$$

where the $\mathbb{A}_n$ entering this formula are the super-amplitudes with stripped-off *bosonic* delta-functions $\delta^{(4)}(p)$. The fermionic delta-functions of Eq. (2.128) remain included. Moreover we have

$$\hat{1}(z_{P_i}) = \{\hat{\lambda}_1(z_{P_i}), \tilde{\lambda}_1, \eta_1\}, \qquad \hat{n}(z_{P_i}) = \{\lambda_1, \hat{\tilde{\lambda}}_n, \hat{\eta}_n\},$$

$$\hat{P}_i(z_{P_i}) = \{\lambda_{\hat{P}_i}, \tilde{\lambda}_{\hat{P}_i}, \eta_{\hat{P}_i}\}, \qquad -\hat{P}_i(z_{P_i}) = \{-\lambda_{\hat{P}_i}, \tilde{\lambda}_{\hat{P}_i}, \eta_{\hat{P}_i}\},$$

$$z_{P_i} = \frac{P_i^2}{\langle n|P_i|1]}, \qquad P_i = p_1 + p_2 + \cdots + p_{i-1}, \qquad (2.145)$$

$$\lambda_{\hat{P}_i}\tilde{\lambda}_{\hat{P}_i} = \hat{\lambda}_1(z)\tilde{\lambda}_1 + \sum_{j=2}^{i-1}\lambda_j\tilde{\lambda}_j = -\sum_{j=i}^{n-1}\lambda_j\tilde{\lambda}_j - \lambda_n\hat{\tilde{\lambda}}_n(z).$$

Note that in the super BCFW-recursion the sum over intermediate states present in the ordinary BCFW-recursion of Eq. (2.16) is now elegantly replaced by an integral over η_{P_i}. What remains to be shown is the vanishing of the residue at $z = \infty$.

The key point here is the observation quoted in Sect. 2.1 based on [4] that a shift with a $(1^+, n^+)$ gluon helicity configurations has a falloff as $1/z$ for $z \to \infty$. How can we relate this to the super-amplitude case? The fact that the super-amplitude is also invariant under the supersymmetry $\bar{q}$, i.e. $\bar{q}_A^{\dot{\alpha}}\mathbb{A}_n = 0$ with $\bar{q}_A^{\dot{\alpha}} = \tilde{\lambda}^{\dot{\alpha}}\partial_{\eta^A}$, may be used to set two arbitrary η_i's to zero. This is seen as follows: a *finite* $\bar{q}$-transformation acts as a translation in η_i-space

$$\eta_i^A \to \eta_i^A + [\xi^A\tilde{\lambda}_i], \qquad (2.146)$$

where $\xi_{\dot{\alpha}}^A$ is the associated Grassmann odd transformation parameter. The special choice

$$\xi_{\dot{\alpha}}^A = \frac{\tilde{\lambda}_{1\dot{\alpha}}\eta_n^A - \tilde{\lambda}_{n\dot{\alpha}}\eta_1^A}{[n1]} \qquad (2.147)$$

sets η_1 and η_n to zero, as $[\xi^A\tilde{\lambda}_{1|n}] = -\eta_{1|n}^A$. In fact the shifts of Eqs. (2.140), (2.143) leave the parameter $\xi_{\dot{\alpha}}^A$ invariant

$$\xi_{\dot{\alpha}}^A(z) = \frac{\tilde{\lambda}_{1\dot{\alpha}}\hat{\eta}_n^A - \hat{\tilde{\lambda}}_{n\dot{\alpha}}\eta_1^A}{[\hat{n}1]} = \frac{\tilde{\lambda}_{1\dot{\alpha}}\eta_n^A - \tilde{\lambda}_{n\dot{\alpha}}\eta_1^A}{[n1]} = \xi_{\dot{\alpha}}^A. \qquad (2.148)$$

Thus by using this finite $\bar{q}$-supersymmetry transformation we can relate the z-dependence of the full superamplitude to that of its component with a $(1^+, n^+)$ positive gluon helicity configuration. The later is known to vanish as $1/z$ for z going to infinity. This then proves the $1/z$ falloff for the full super-amplitude and the vanishing of the residuum at infinity of Eq. (2.144) for any two neighboring legs in $\mathbb{A}_n(1, \ldots, n)$.

2.7.5 Three-Point Super-amplitudes

As before the seed for solving the recursion are the 3-point super-amplitudes. As in the pure Yang-Mills case we have a MHV_3 and $\overline{\text{MHV}}_3$ super-amplitude characterized by either $[ij] = 0$ or $\langle ij \rangle = 0$ for $i, j = 1, 2, 3$, respectively.

Up to a proportionality constant the 3-point MHV super-amplitude is uniquely determined. It must have the form $\mathbb{A}_3^{\text{MHV}} \sim \delta^{(4)}(p)\delta^{(8)}(q)$. The 3-point kinematics $[ij] = 0$ but $\langle ij \rangle \neq 0 \; \forall i, j \in \{1, 2, 3\}$ along with the local helicity requirement $h_i \circ \mathbb{A}_3 = \mathbb{A}_3$, $\forall i$, uniquely leads via Eq. (2.124) to the result

$$\mathbb{A}_3^{\text{MHV}} = \frac{\delta^{(4)}(p)\delta^{(8)}(q)}{\langle 12 \rangle \langle 23 \rangle \langle 31 \rangle}. \tag{2.149}$$

Its cousin, the $\overline{\text{MHV}}_3$ super-amplitude follows from parity and a Fourier transformation in $\bar{\eta}$-space. This is seen as follows: conjugating the MHV amplitude Eq. (2.149) we have

$$\overline{\mathbb{A}_3^{\text{MHV}}} = -\frac{\delta^{(4)}(p)}{[12][23][31]}\delta^{(8)}\left(\sum_{i=1}^{3} \tilde{\lambda}_i \bar{\eta}_i\right). \tag{2.150}$$

As we are dealing with a chiral on-shell superspace initially $\{\lambda_i, \tilde{\lambda}_i, \eta_i\}$ we need to Fourier transform this result back to η-space, i.e. we have the relation

$$\mathbb{A}_3^{\overline{\text{MHV}}} = \int \left(\prod_{i=1}^{3} d^4\bar{\eta}_i\right) e^{i \sum_{i=1}^{3} \eta_i^A \bar{\eta}_{iA}} \overline{\mathbb{A}_3^{\text{MHV}}}. \tag{2.151}$$

It is instructive to perform this integral.

For this we need the following identity (to be shown in Exercise 2.9)

$$\delta^{(2)}\left(\lambda^\alpha a + \mu^\alpha b\right) = \begin{cases} \frac{\delta(a)\delta(b)}{|\langle \lambda\mu \rangle|} & \text{for } a, b \text{ Grassmann even,} \\ \delta(a)\delta(b)\langle \lambda\mu \rangle & \text{for } a, b \text{ Grassmann odd} \end{cases} \tag{2.152}$$

which is a consequence of the linear independence of the two-vectors λ^α and μ^α. Using this one shows that

$$\delta^{(8)}\left(\sum_{i=1}^{3} \tilde{\lambda}_i^{\dot{\alpha}} \bar{\eta}_{iA}\right) = [12]^4 \delta^{(4)}\left(\bar{\eta}_{1,A} - \frac{[23]}{[12]}\bar{\eta}_{3A}\right)\delta^{(4)}\left(\bar{\eta}_{2,A} - \frac{[31]}{[12]}\bar{\eta}_{3A}\right). \tag{2.153}$$

Inserting this expression in Eq. (2.151) and performing the $\bar{\eta}_1$ and $\bar{\eta}_2$ integrals via the delta functions yields

$$\int \left(\prod_{i=1}^{3} d^4 \bar{\eta}_i\right) \delta^{(8)} \left(\sum_{i=1}^{3} \tilde{\lambda}_i^{\dot{\alpha}} \bar{\eta}_{iA}\right) = [12]^4 \int d^4 \bar{\eta}_3 e^{i\bar{\eta}_{3A}(\eta_3^A + \frac{[23]}{[12]}\eta_1^A + \frac{[31]}{[12]}\eta_2^A)}$$

$$= [12]^4 \prod_{A=1}^{4} \delta\left(\eta_3^A + \frac{[23]}{[12]}\eta_1^A + \frac{[31]}{[12]}\eta_2^A\right)$$

$$= \delta^{(4)}\left([12]\eta_3^A + [23]\eta_1^A + [31]\eta_2^A\right), \quad (2.154)$$

where we have also used the integral representation for the Grassmann odd delta function of Eq. (2.132). In summary, we find the final expression for the $\overline{\text{MHV}}_3$ amplitude

$$\mathbb{A}_3^{\overline{\text{MHV}}} = -\frac{\delta^{(4)}(p)\delta^{(4)}([12]\eta_3 + [23]\eta_1 + [31]\eta_2)}{[12][23][31]}. \quad (2.155)$$

At first sight it is surprising that for this amplitude the invariance under q supersymmetry does not require the $\delta^{(8)}(q)$ factor as claimed in Eq. (2.128) for $n > 3$. The reason lies in the special kinematics of the $\overline{\text{MHV}}_3$ amplitude where $\langle ij \rangle = 0$ requires all λ_i to be parallel. Therefore q factorizes into $q^{\alpha A} = \lambda_F^\alpha \eta_F^A$ for suitable λ_F and η_F. Hence q-invariance only demands a factor of $\delta^{(4)}(\eta_F)$ which is what we indeed found in Eq. (2.155).

Exercise 2.9 (Super Delta Function Manipulations)

   Prove the following relations for the helicity spinors λ and μ

$$\delta^{(2)}\left(\lambda^\alpha a + \mu^\alpha b\right) = \begin{cases} \frac{\delta(a)\delta(b)}{|\langle\lambda\mu\rangle|} & \text{for } a, b \text{ Grassmann even,} \\ \delta(a)\delta(b)\langle\lambda\mu\rangle & \text{for } a, b \text{ Grassmann odd.} \end{cases}$$

Use this to show that

$$\delta^{(8)}\left(\sum_{i=1}^{3} \lambda_i^\alpha \eta_i^A\right) = \langle12\rangle^4 \delta^{(4)}\left(\eta_1^A - \frac{\langle23\rangle}{\langle12\rangle}\eta_3^A\right) \delta^{(4)}\left(\eta_2^A - \frac{\langle31\rangle}{\langle12\rangle}\eta_3^A\right),$$

which we used in its complex conjugated version in the derivation of the $\mathbb{A}_3^{\overline{\text{MHV}}}$ amplitude above.

2.7.6 Solving the Super-BCFW Recursion: MHV Case

If one decomposes the super-BCFW recursion into contributions of different Grassmann degrees, i.e. decomposes the super-amplitudes into their various N^PMHV

Fig. 2.9 Classification of amplitudes according to helicity violating level k and number of external legs n. *Every black dot denotes an* $N^k MHV_n$ (super-)amplitude. The three-point amplitudes are special. The recursion relations can be solved iteratively in n and k, as explained in the main text

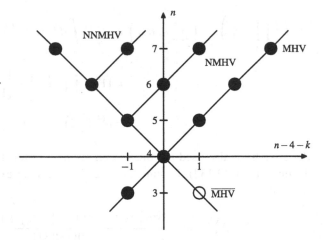

parts as in Eq. (2.129), one obtains a recursion of the form

$$\mathbb{A}_n^{N^p MHV} = \int \frac{d^4 \eta_P}{P^2} \mathbb{A}_3^{\overline{MHV}}(z_P) \mathbb{A}_{n-1}^{N^p MHV}(z_P)$$

$$+ \sum_{m=0}^{p-1} \sum_{i=4}^{n-1} \int \frac{d^4 \eta_{P_i}}{P_i^2} \mathbb{A}_i^{N^m MHV}(z_{P_i}) \mathbb{A}_{n-i+2}^{N^{(p-m-1)} MHV}(z_{P_i}). \quad (2.156)$$

The reason for this decomposition lies in the fact that the total η-count on the left-hand-side, η^{4p+8}, has to equal the η-count on the right-hand-side. Due to the Grassmann integral $\int d^4 \eta_P$ the combined η-count of the integrand $\mathbb{A}^L \cdot \mathbb{A}^R$ has to be by four larger than the left-hand-side. Note that we have not included a term where the left subamplitude is $\mathbb{A}_3^{MHV}$. This cannot happen since $\mathbb{A}_3^{MHV}(1, 2, *)$ implies the vanishing of the square bracket [12]. In the left subamplitude $\tilde{\lambda}_1$ is unshifted and hence $[12] = 0$ implies $(p_1 + p_2)^2 = 0$, which is a restriction on the kinematics not generally true. In consequence such a term does not contribute to the recursion relation as there is no solution to the on-shell conditions for this channel for generic external momenta. Similarly, the right sub-amplitude may never be $\mathbb{A}_3^{\overline{MHV}}$.

This form of the super-BCFW recursion (2.156) leads to the following iterative structure: the MHV_n super-amplitudes follow from the lower-point $MHV_{k<n}$ and the $\overline{MHV}_3$ super-amplitudes. The $NMHV_n$ super-amplitudes arise from the $NMHV_{n-1}$, $MHV_{k<n}$ and $\overline{MHV}_3$. An iterative solution of the $NNMHV$ super-amplitudes requires the knowledge of the $NMHV$, MHV and $\overline{MHV}_3$ superamplitudes etc. See Fig. 2.9 for illustration.

To start this iteration we look at the MHV sector. We already know the form of the MHV super-amplitude by lifting the result of the MHV amplitude at component level to superspace

$$\mathbb{A}_n^{MHV}(\lambda_i, \tilde{\lambda}_i, \eta_i) = \frac{\delta^{(4)}(p)\delta^{(8)}(q)}{\langle 12 \rangle \langle 23 \rangle \cdots \langle n1 \rangle}. \quad (2.157)$$

Let us nevertheless rederive this formula from the super-BCFW recursion for the 4-point case. For $n = 4$ and $p = 0$ Eq. (2.156) only has one term and simply reads

$$
\mathbb{A}_4^{\text{MHV}} = \int \frac{d^4 \eta_P}{P^2} \mathbb{A}_3^{\overline{\text{MHV}}}(z_P) \mathbb{A}_3^{\text{MHV}}(z_P)
$$

$$
= -\int \frac{d^4 \eta_P}{P^2} \frac{\delta^{(4)}(\eta_1[2, \hat{P}] + \eta_2[\hat{P}1] + \eta_P[12]) \delta^{(8)}(\lambda_{\hat{P}} \eta_P + \lambda_3 \eta_3 + \lambda_4 \hat{\eta}_4)}{[12][2\hat{P}][\hat{P}1]\langle\hat{P}3\rangle\langle34\rangle\langle4\hat{P}\rangle}.
$$

(2.158)

We use the $\delta^{(4)}$-function to localize η_P

$$
\eta_P = -\frac{1}{[12]}\left(\eta_1[2\hat{P}] + \eta_2[\hat{P}1]\right). \tag{2.159}
$$

Inserting this into the $\delta^{(8)}$-function and using momentum conservation yields

$$
\delta^{(8)}\left(-\frac{\lambda_{\hat{P}}}{[12]}\left(\eta_1[2\hat{P}] + \eta_2[\hat{P}1]\right) + \lambda_3 \eta_3 + \lambda_4 \hat{\eta}_4\right)
$$

$$
\stackrel{\lambda_{\hat{P}}\tilde{\lambda}_{\hat{P}} = \hat{\lambda}_1\tilde{\lambda}_1 + \lambda_2\tilde{\lambda}_2}{=} \delta^{(8)}\left(-\frac{\hat{\lambda}_1\eta_1[21] + \lambda_2\eta_2[21]}{[12]} + \lambda_3\eta_3 + \lambda_4\hat{\eta}_4\right) = \delta^{(8)}(q),
$$

(2.160)

recovering the expected q-preserving delta-function. Note that in the last step there is a cancelation of the z_P dependent terms in $\hat{\lambda}_1 \eta_1 + \lambda_4 \hat{\eta}_4 = \lambda_1 \eta_1 + \lambda_4 \eta_4$. In the following, we will often perform similar calculations using the Grassmann delta functions. In this way, the intermediate state sums, which in this formalism are given by Grassmann integrals, can be trivially performed, and this considerably streamlines the computations. All that remains are the bosonic factors which we group as follows

$$
-\frac{1}{P^2}[12]^4 \frac{1}{[12]\langle34\rangle}\frac{1}{[2\hat{P}]\langle4\hat{P}\rangle}\frac{1}{[\hat{P}1]\langle\hat{P}3\rangle}. \tag{2.161}
$$

Here the second term arises from pulling the factor $[12]$ out of the $\delta^{(4)}$-fermionic delta function. Using momentum conservation $|\hat{P}\rangle\langle\hat{P}| = |1]\langle\hat{1}| + |2]\langle2|$ we have

$$
[2\hat{P}]\langle4\hat{P}\rangle = [21]\langle4\hat{1}\rangle = [21]\langle41\rangle, \qquad [\hat{P}1]\langle\hat{P}3\rangle = [21]\langle23\rangle, \qquad P^2 = \langle12\rangle[21],
$$

(2.162)

which enables one to bring the expression in (2.161) into the form

$$
-\frac{1}{P^2}[12]^4 \frac{1}{[12]\langle34\rangle}\frac{1}{[2\hat{P}]\langle4\hat{P}\rangle}\frac{1}{[\hat{P}1]\langle\hat{P}3\rangle} = \frac{1}{\langle12\rangle\langle23\rangle\langle34\rangle\langle41\rangle}. \tag{2.163}
$$

This proves Eq. (2.157) for $n = 4$. The proof for general n-point MHV superamplitudes works analogously.

2.7.7 Solving the Super-BCFW Recursion: NMHV Case

Finally let us solve the recursion for the NMHV case closely following the approach of [9]. The general recursion formula Eq. (2.156) for $p = 1$ reads

$$
\mathbb{A}_n^{\text{NMHV}} = \int \frac{d^4 P}{P^2} \int d^4 \eta_{\hat{P}} \mathbb{A}_3^{\overline{\text{MHV}}}(z_P) \mathbb{A}_{n-1}^{\text{NMHV}}(z_P)
$$

$$
+ \sum_{i=4}^{n-1} \int \frac{d^4 P_i}{P_i^2} \int d^4 \eta_{P_i} \mathbb{A}_i^{\text{MHV}}(z_{P_i}) \mathbb{A}_{n-i+2}^{\text{MHV}}(z_{P_i})
$$

$$
\equiv A + B, \tag{2.164}
$$

resulting in a homogeneous term A and an inhomogeneous term B.

The inhomogeneous term may be straightforwardly computed from the known MHV amplitudes. Writing the Grassmann delta function coming from the left $\mathbb{A}_i^{\text{MHV}}(z_P)$ in the following way,

$$
\delta^{(8)}\left(\hat{\lambda}_1 \eta_1 + \sum_{j=2}^{i-1} \lambda_j \eta_j - \lambda_{\hat{P}_i} \eta_{P_i}\right)
$$

$$
= \langle \hat{1}\hat{P}_i \rangle^4 \delta^{(4)}\left(\sum_{j=2}^{i-1} \frac{\langle \hat{1}j \rangle}{\langle \hat{1}\hat{P}_i \rangle} \eta_j - \eta_{P_i}\right) \delta^{(4)}\left(\eta_1 + \sum_{j=2}^{i-1} \frac{\langle j\hat{P}_i \rangle}{\langle \hat{1}\hat{P}_i \rangle} \eta_j\right), \tag{2.165}
$$

the integration over η_{P_i} may be performed straightforwardly. In this way, we obtain the following contribution to the n-point NMHV amplitude:

$$
B = \frac{\delta^{(4)}(p)\delta^{(8)}(q)}{\prod_{j=1}^n \langle jj+1 \rangle} \sum_{i=4}^{n-1} R_{n;2i}. \tag{2.166}
$$

Here $R_{r;st}$ (called R-invariant because of its properties under dual superconformal symmetry, cf. Chap. 4) is given by

$$
R_{r;st} = \frac{\langle ss-1 \rangle \langle tt-1 \rangle \delta^{(4)}(\Xi_{r;st})}{x_{st}^2 \langle r|x_{rs}x_{st}|t \rangle \langle r|x_{rs}x_{st}|t-1 \rangle \langle r|x_{rt}x_{ts}|s \rangle \langle r|x_{rt}x_{ts}|s-1 \rangle}, \tag{2.167}
$$

where

$$
x_{ab} := p_a + p_{a+1} + \cdots + p_{b-1}, \qquad \theta_{ab} := q_a + q_{a+1} + \cdots + q_{b-1}, \tag{2.168}
$$

are the dual variables or region momenta which will play a more prominent rôle in chapter four. The Grassmann odd quantity $\Xi_{r;st}$ in the above is given by

$$
\Xi_{r;st} = \langle r|x_{rs}x_{st}|\theta_{tr} \rangle + \langle r|x_{rt}x_{ts}|\theta_{sr} \rangle. \tag{2.169}
$$

In the following we will often deal with the quantity $\Xi_{n;st}$ for $1 < s < t < n$. It is instructive to switch from the dual θ_i to the η_i,

$$
\Xi_{n;st} = \langle n | \left[x_{ns} x_{st} \sum_{i=t}^{n-1} |i\rangle \eta_i + x_{nt} x_{ts} \sum_{i=s}^{n-1} |i\rangle \eta_i \right], \tag{2.170}
$$

to see that $\Xi_{n;st}$ is independent of η_n and η_1. Alternatively, using the $\delta^{(8)}(q)$ present in all physical amplitudes to rewrite the sums we can obtain

$$
\delta^{(8)}(q) \Xi_{n;st} = -\delta^{(8)}(q) \langle n | \left[x_{ns} x_{st} \sum_{i=1}^{t-1} |i\rangle \eta_i + x_{nt} x_{ts} \sum_{i=1}^{s-1} |i\rangle \eta_i \right], \tag{2.171}
$$

such that the only dependence on η_{n-1} and η_n on the l.h.s. of (2.171) is contained in $\delta^{(8)}(q)$.

Moreover, it is useful to realize that terms like $\langle r | x_{rs} x_{st} | t \rangle$ in (2.167) and similar terms in (2.169) can always be written as

$$
\langle r | x_{rs} x_{st} | t \rangle = \langle r | x_{r+1s} x_{st} | t \rangle, \tag{2.172}
$$

such that it is clear that they only depend explicitly on λ_r, but not on $\tilde{\lambda}_r$.

Finally note that the superamplitude must have cyclic symmetry. This implies

$$
\delta^{(8)}(q) R_{5;24} = \delta^{(8)}(q) R_{1;35} = \delta^{(8)}(q) R_{2;41} = \delta^{(8)}(q) R_{3;52} = \delta^{(8)}(q) R_{4;13}. \tag{2.173}
$$

This is just the first example of a general identity for n points

$$
\delta^{(8)}(q) \sum_{s,t} R_{r;st} = \delta^{(8)}(q) \sum_{s,t} R_{r';st}, \tag{2.174}
$$

where the sum goes over all values of s, t such that r, s, t (or r', s, t) are ordered cyclically with r and s (or r' and s) and s and t separated by at least two.

Let us first analyze the 5-point NMHV amplitude. This is a somewhat trivial case as for five points, NMHV$_5$ = $\overline{\text{MHV}}_5$, and therefore we could simply obtain the NMHV$_5$ amplitude from a Grassmann Fourier transform of the known $\overline{\text{MHV}}_5$ amplitude similar to the way we obtained the $\overline{\text{MHV}}_3$ amplitude above. Nevertheless it is instructive to evaluate the recursion (2.164) in this case.

One immediately concludes that only the second term in (2.164) contributes, because there is no four-point NMHV amplitude. Hence for five points, the complete amplitude is given by the inhomogeneous term (2.166), i.e.

$$
\mathbb{A}_5^{\text{NMHV}} = \frac{\delta^{(4)}(p) \delta^{(8)}(q)}{\prod_{j=1}^{5} \langle jj+1 \rangle} R_{5;24}. \tag{2.175}
$$

Moving on to the general NMHV$_n$ case it can be seen that there is a general pattern of how the n-point solution is generated from the $(n-1)$-point one. We therefore postulate that the ansatz

$$\mathbb{A}_n^{\text{NMHV}} = \mathbb{A}_n^{\text{MHV}} \mathscr{P}_n^{\text{NMHV}} = \frac{\delta^{(4)}(p)\delta^{(8)}(q)}{\langle 12 \rangle \langle 23 \rangle \cdots \langle n1 \rangle} \sum_{2 \leq s < t \leq n-1} R_{n;st}, \qquad (2.176)$$

solves the supersymmetric BCFW-recursion. In this expression it is assumed that s and t are separated by at least two. We can verify that Eq. (2.176) is correct for $n = 5$ by comparing to Eq. (2.175).

We now prove (2.176) by induction. Let us assume that the form (2.176) is valid for $n - 1$ points. Then it follows from the cyclicity of superamplitudes that (2.174) is also true for $n - 1$ points. Now, we notice that $\mathbb{A}_{n-1}^{\text{NMHV}}(z_P)$ in the homogeneous term, A on the RHS of (2.164), only involves the quantities $R_{n-1;st}$ where the first subscript is always equal to $n - 1$. Cyclic symmetry allows us to insert $\mathbb{A}_{n-1}^{\text{NMHV}}(z_P)$ into (2.164) in our favorite orientation. It is convenient to insert it such that the legs $\{1, 2, 3, \ldots, n-1\}$ of $\mathbb{A}_{n-1}^{\text{NMHV}}(z_P)$ are identified with the legs $\{\hat{P}, 3, 4, \ldots, n\}$ in the recursion relation

$$A = \int \frac{d^4 P}{P^2} \int d^4 \eta_{\hat{P}} \mathbb{A}_3^{\overline{\text{MHV}}}(z_P) \mathbb{A}_{n-1}^{\text{MHV}} \mathscr{P}_{n-1}(\hat{P}, 3, \ldots, \bar{n}). \qquad (2.177)$$

After carrying out this change of labels in $\mathbb{A}_{n-1}^{\text{NMHV}}(z_P)$ is clear from Eqs. (2.170) and (2.172) that the obtained $R_{n;st}$ does not depend on $\eta_{\hat{P}}$. Indeed the range of η-dependence is only $\{\eta_3, \ldots, \eta_{n-1}\}$. When the lower summation variable attains its minimum value, there is an explicit dependence on the spinor $\langle \hat{P}|$. However, due to the three-point kinematics, this spinor is proportional to $\langle 2|$ and since it appears homogeneously in R with degree zero it can simply be replaced by $\langle 2|$. Thus we find

$$A = \frac{\delta^{(4)}(p)\delta^{(8)}(q)}{\prod_{j=1}^{n} \langle jj+1 \rangle} \sum_{3 \leq s < t \leq n-1} R_{n;st}. \qquad (2.178)$$

We see that (2.166) is just the missing first term (for $s = 2$) to complete (2.178) to the ansatz (2.176) for n points, i.e.

$$A + B = \mathbb{A}_n^{\text{NMHV}} = \frac{\delta^{(4)}(p)\delta^{(8)}(q)}{\prod_{j=1}^{n} \langle jj+1 \rangle} \sum_{2 \leq s < t \leq n-1} R_{n;st}. \qquad (2.179)$$

This completes the inductive proof for the general NMHV super-amplitude.

In fact it is possible to *completely solve* the super-BCFW recursion and write down an exact analytic expression of all tree super-amplitudes in $\mathcal{N} = 4$ super Yang-Mills [9].

Exercise 2.10 (The n-Point MHV Superamplitude and Component Amplitudes)
Use the super-BCFW recursion to prove the MHV super-amplitude formula at n-points. Use this to establish the four point gluino-quark component field amplitudes

$$A_4\left(1_{\bar{g}}^-, 2_{\bar{g}}^+, 3^-, 4^+\right) = \delta^{(4)}(p)\frac{\langle 31 \rangle^3 \langle 23 \rangle}{\langle 12 \rangle \cdots \langle n1 \rangle},$$

$$A_4\left(1_{\bar{g}}^-, 2_{\bar{g}}^+, 3_{\bar{g}}^-, 4_{\bar{g}}^+\right) = -\delta^{(4)}(p)\frac{\langle 31 \rangle^3 \langle 24 \rangle}{\langle 12 \rangle \cdots \langle n1 \rangle}. \qquad (2.180)$$

What follows from this result for the 4-point single-flavor massless QCD tree-level amplitudes with one and two quark lines?

Exercise 2.11 (Extraction of Split-Helicity Gluon Amplitudes from NMHV Super-amplitude)

Start from the formula for the n-point NMHV superamplitude. As a component in its η expansion, it contains all pure Yang-Mills gluon amplitudes with 3 minus-helicity and $(n - 3)$ plus-helicity gluons. Here we focus on the so-called split-helicity gluon amplitudes, where the negative helicity gluons are adjacent. It is found in the superamplitude in the following way,

$$\mathcal{A}_n^{\text{NMHV}} = (\eta_{n-2})^4 (\eta_{n-1})^4 (\eta_n)^4$$
$$\times A\left(1^+, \ldots, (n-3)^+, (n-2)^-, (n-1)^-, n^-\right) + \cdots, \qquad (2.181)$$

where the dots denote other terms in the η expansion. A convenient way to project out the amplitude that we are interested in is

$$A\left(1^+, \ldots, (n-3)^+, (n-2)^-, (n-1)^-, n^-\right)$$
$$= \int d^4\eta_{n-2} \int d^4\eta_{n-1} \int d^4\eta_n \, \mathcal{A}_n^{\text{NMHV}}. \qquad (2.182)$$

Carry out the Grassmann integrals in (2.182).

- Hint 1: Write the super momentum conserving Grassmann delta function as

$$\delta^{(8)}\left(q_\alpha^A\right) = \langle n-1 n \rangle^4 \delta^{(4)}\left(\eta_{n-1}^A + \sum_{i=1}^{n-2} \frac{\langle in \rangle}{\langle n-1 n \rangle} \eta_i^A\right)$$

$$\times \delta^{(4)}\left(\eta_n^A + \sum_{i=1}^{n-2} \frac{\langle n-1 i \rangle}{\langle n-1 n \rangle} \eta_i^A\right). \qquad (2.183)$$

- Hint 2: Use the cyclic symmetry of the NMHV superamplitude in order to write it in a form where the dependence on the relevant η's is simple. Different choices of the representation of $\mathcal{A}_n^{\text{NMHV}}$ will lead to equivalent answers, but may differ e.g. in the number of terms.

The solution to this exercise can be found in Sect. 8 of Ref. [9].

2.8 References and Further Reading

In Sect. 2.1 we introduced recursion relations enabling the construction of higher-point tree-level amplitudes from lower-point ones using exclusively on-shell data. The described formalism based on a two leg shift is due to Britto, Cachazo, Feng and Witten from 2005 [2, 10]. It is the most efficient technique in a number of recursion relation families.

The factorization properties of scattering amplitudes into sub-amplitudes discussed in Sect. 2.4 are a direct consequence of the color-ordered Feynman rules. The universality of the soft limit was shown by Berends and Giele [11] based on their off-shell recursion relation [12] discussed below. The universality of the factorization on multi-particle poles and collinear limits may also be explained via the point-like limit of string theory amplitudes, see [1].

Our discussion of the conformal invariance of gluon scattering amplitudes of Sect. 2.5 follows the paper of Witten [13].

The $\mathcal{N} = 4$ super Yang-Mills (SYM) theory introduced in Sect. 2.6 was first written down in 1976 [14, 15], its finiteness properties were analyzed in the early 1980s [16–18] and the concept of super-amplitudes in an on-shell chiral superspace goes back to Nair [19] who wrote down the MHV n-point super amplitudes. The supersymmetrized version of the BCFW recursion discussed in Chap. 2.6 was derived in [20] and a related construction was given in [21, 22]. The complete analytic solution of the super BCFW recursion was achieved in 2008 in [9] giving compact analytical formulae for *all* tree-amplitudes in $\mathcal{N} = 4$ SYM. This analytic solution was subsequently used to generate *all* tree-level QCD amplitudes with up to four massless quark-anti-quark pairs and an arbitrary number of gluons [23] including an implementation of the formulas in Mathematica.

In fact, recursive techniques for the construction of tree-level amplitudes existed before the BCFW-recursion discussed in these notes. In 1988 Berends and Giele introduced an off-shell recursion [12] using as building blocks color ordered amplitudes with one off-shell leg. This method is also highly efficient and easily implemented in numerical applications. For a pedagogical introduction to the method see e.g. Chap. 3 of [24].

An alternative and purely on-shell recursion in gauge theories is the Cachazo, Svrcek and Witten MHV vertex expansion from 2004 [25]. This CSW or MHV vertex expansion uses as building blocks exclusively MHV n-point amplitudes. Here all N^PMHV amplitudes may be represented as on-shell diagrams with MHV amplitudes as vertices. Although it was historically not developed in this way, it may be viewed as a particular form of the BCFW recursions, where one shifts several external momenta in a specific way [26]. See also [3] for the supersymmetric case.

Furthermore a series of papers established a Lagrangian formulation of the MHV-vertex expansion [27–31]. Here field redefinitions and suitable gauge choices reformulate the (super) Yang-Mills theory as a model with MHV vertices of arbitrary multiplicity whose Feynman rules yield the CSW expansion. A good overview to the MHV vertex expansion at tree and loop-level is given in [32].

The recursive techniques discussed here can be generalized to scattering amplitudes with massive particles as we briefly discussed. Here the spinor helicity formalism may be extended to massive momenta by representing them as the sum of two null momenta. Concretely one introduces a reference null-momenta q and writes $p_i^\mu = p_{\perp i}^\mu + \frac{m_i^2}{2q \cdot p_i} q^\mu$ with a null $p_{\perp i}^\mu$. This decomposition ensures $p_i^2 = m_i^2$, for a full exposition of this construction see [33]. The analogue of the BCFW recursion relation for amplitudes with massive particles was discussed in [7, 34]. An analogue of the CSW vertex expansion also exists in the massive case [35, 36].

Finally, in these notes we have focused on pure Yang-Mills and on maximally supersymmetric Yang-Mills theory. A detailed account of on-shell superspace techniques for theories with $\mathcal{N} < 4$ supersymmetry is given in [37].

References

1. M.L. Mangano, S.J. Parke, Multi-parton amplitudes in gauge theories. Phys. Rep. **200**, 301–367 (1991). arXiv:hep-th/0509223
2. R. Britto, F. Cachazo, B. Feng, E. Witten, Direct proof of tree-level recursion relation in Yang-Mills theory. Phys. Rev. Lett. **94**, 181602 (2005). arXiv:hep-th/0501052
3. H. Elvang, D.Z. Freedman, M. Kiermaier, Proof of the MHV vertex expansion for all tree amplitudes in $N = 4$ SYM theory. J. High Energy Phys. **0906**, 068 (2009). arXiv:0811.3624
4. N. Arkani-Hamed, J. Kaplan, On tree amplitudes in gauge theory and gravity. J. High Energy Phys. **0804**, 076 (2008). arXiv:0801.2385
5. N. Arkani-Hamed, F. Cachazo, J. Kaplan, What is the simplest quantum field theory? (2008). arXiv:0808.1446
6. Z. Bern, L.J. Dixon, D.A. Kosower, Two-loop g-gt; gg splitting amplitudes in QCD. J. High Energy Phys. **0408**, 012 (2004). arXiv:hep-ph/0404293
7. S.D. Badger, E.W.N. Glover, V.V. Khoze, P. Svrcek, Recursion relations for gauge theory amplitudes with massive particles. J. High Energy Phys. **0507**, 025 (2005)
8. M.F. Sohnius, Introducing supersymmetry. Phys. Rep. **128**, 39–204 (1985)
9. J.M. Drummond, J.M. Henn, All tree-level amplitudes in $\mathcal{N} = 4$ SYM. J. High Energy Phys. **0904**, 018 (2009). arXiv:0808.2475
10. R. Britto, F. Cachazo, B. Feng, New recursion relations for tree amplitudes of gluons. Nucl. Phys. B **715**, 499 (2005). arXiv:0808.2475
11. F.A. Berends, W.T. Giele, Multiple soft gluon radiation in parton processes. Nucl. Phys. B **313**, 595 (1989)
12. F.A. Berends, W.T. Giele, Recursive calculations for processes with n gluons. Nucl. Phys. B **306**, 759 (1988)
13. E. Witten, Perturbative gauge theory as a string theory in twistor space. Commun. Math. Phys. **252**, 189 (2004). arXiv:hep-th/0312171
14. F. Gliozzi, J. Scherk, D.I. Olive, Supersymmetry, supergravity theories and the dual spinor model. Nucl. Phys. B **122**, 253–290 (1977)
15. L. Brink, J.H. Schwarz, J. Scherk, Supersymmetric Yang-Mills theories. Nucl. Phys. B **121**, 77 (1977)
16. M.F. Sohnius, P.C. West, Conformal invariance in $\mathcal{N} = 4$ supersymmetric Yang-Mills theory. Phys. Lett. B **100**, 245 (1981)
17. P.S. Howe, K.S. Stelle, P.K. Townsend, Miraculous ultraviolet cancellations in supersymmetry made manifest. Nucl. Phys. B **236**, 125 (1984)
18. L. Brink, O. Lindgren, B.E.W. Nilsson, $N = 4$ Yang-Mills theory on the light cone. Nucl. Phys. B **212**, 401 (1983)

19. V.P. Nair, A current algebra for some gauge theory amplitudes. Phys. Lett. B **214**, 215 (1988)
20. N. Arkani-Hamed, F. Cachazo, C. Cheung, J. Kaplan, A duality for the S matrix. J. High Energy Phys. **03**, 020 (2010)
21. A. Brandhuber, P. Heslop, G. Travaglini, A note on dual superconformal symmetry of the $\mathcal{N} = 4$ super Yang-Mills S-matrix. Phys. Rev. D **78**, 125005 (2008). arXiv:0807.4097
22. M. Bianchi, H. Elvang, D.Z. Freedman, Generating tree amplitudes in $\mathcal{N} = 4$ SYM and $\mathcal{N} = 8$ SG. J. High Energy Phys. **09**, 063 (2008). arXiv:0805.0757
23. L.J. Dixon, J.M. Henn, J. Plefka, T. Schuster, All tree-level amplitudes in massless QCD. J. High Energy Phys. **1101**, 035 (2011). arXiv:1010.3991
24. L.J. Dixon, Calculating scattering amplitudes efficiently (1996). arXiv:hep-ph/9601359
25. F. Cachazo, P. Svrcek, E. Witten, MHV vertices and tree amplitudes in gauge theory. J. High Energy Phys. **09**, 006 (2004). arXiv:hep-th/0403047
26. K. Risager, A direct proof of the CSW rules. J. High Energy Phys. **0512**, 003 (2005). arXiv:hep-th/0508206
27. A. Gorsky, A. Rosly, From Yang-Mills Lagrangian to MHV diagrams. J. High Energy Phys. **0601**, 101 (2006). arXiv:hep-th/0510111
28. P. Mansfield, The Lagrangian origin of MHV rules. J. High Energy Phys. **0603**, 037 (2006). arXiv:hep-th/0511264
29. J.H. Ettle, C.-H. Fu, J.P. Fudger, P.R.W. Mansfield, T.R. Morris, S-matrix equivalence theorem evasion and dimensional regularisation with the canonical MHV Lagrangian. J. High Energy Phys. **0705**, 011 (2007). arXiv:hep-th/0703286
30. H. Feng, Y.-t. Huang, MHV Lagrangian for $N = 4$ super Yang-Mills. J. High Energy Phys. **0904**, 047 (2009). arXiv:hep-th/0611164
31. R. Boels, L.J. Mason, D. Skinner, From twistor actions to MHV diagrams. Phys. Lett. B **648**, 90–96 (2007). arXiv:hep-th/0702035
32. A. Brandhuber, B. Spence, G. Travaglini, Tree-level formalism. J. Phys. A **44**, 454002 (2011). arXiv:1103.3477
33. S. Dittmaier, Weyl-van der Waerden formalism for helicity amplitudes of massive particles. Phys. Rev. D **59**, 016007 (1998). arXiv:hep-ph/9805445
34. S.D. Badger, E.W.N. Glover, V.V. Khoze, Recursion relations for gauge theory amplitudes with massive vector bosons and fermions. J. High Energy Phys. **0601**, 066 (2006). arXiv:hep-th/0507161
35. R. Boels, C. Schwinn, CSW rules for a massive scalar. Phys. Lett. B **662**, 80 (2008). arXiv:0712.3409
36. R. Boels, C. Schwinn, CSW rules for massive matter legs and glue loops. Nucl. Phys., Proc. Suppl. **183**, 137–142 (2008). arXiv:0805.4577
37. H. Elvang, Y.-t. Huang, C. Peng, On-shell superamplitudes in $N = 4$ SYM. J. High Energy Phys. **1109**, 031 (2011). arXiv:1102.4843

Chapter 3
Loop-Level Structure

In this chapter we discuss various aspects of loop-level scattering amplitudes. We begin by defining our conventions for Minkowski-space loop integrals, and explain how they are defined for general dimension D. We then briefly explain the origin of ultraviolet and infrared divergences. Next, we discuss the regularization scheme we use in the remainder of this chapter. We then discuss, at the one-loop level, how to reduce generic loop integrals that appear in Feynman diagram calculations to a basis of scalar integrals. The following section is devoted to a modern technique for the determination of one-loop amplitudes known as generalized unitarity. Here the philosophy is similar to our discussion in the previous chapter in that on-shell tree-level amplitudes are used to construct higher-loop ones. The final sections of this chapter are devoted to methods for evaluating scalar integrals at one loop and beyond, such as Mellin-Barnes transformations, integration by parts and the method of differential equations.

3.1 Introduction

In order to compare theory predictions with measurements from particle colliders it is important to reduce the theoretical uncertainties and scale dependence as far as possible. It is therefore desirable to compute perturbative corrections to the tree-level processes considered in the previous chapter.

In doing so, one encounters loop diagrams, where one has to integrate over the loop momenta in four-dimensional Minkowski space.

There are at least two new effects at loop level that represent a complication with respect to the analysis at tree-level. The first are ultraviolet (UV) divergences, which arise in general due to the bad high energy behavior of the loop integrals. They are absorbed through a renormalization of the bare parameters appearing in the Lagrangian. The standard way to deal with these divergences is by defining the theory in $D = 4 - 2\varepsilon$ dimensions, with $\varepsilon > 0$. This step is called dimensional regularization. Then, after the appropriate parameters in the Lagrangian are renormalized, one can take the limit $\varepsilon \to 0$.

J.M. Henn, J.C. Plefka, *Scattering Amplitudes in Gauge Theories*,
Lecture Notes in Physics 883, DOI 10.1007/978-3-642-54022-6_3,
© Springer-Verlag Berlin Heidelberg 2014

A second complication has to do with specific divergences that can arise when massless particles are present in on-shell processes. They are infrared (IR) divergences. As we will see, these occur due to specific regions in the loop integration. There are two types of such regions, which partially overlap: one is the vanishing energy of the virtual particle in the loop, the second are loop momentum configurations collinear to external momenta. We refer to both as IR divergences. These divergences can also be dealt with in the framework of dimensional regularization. After the ultraviolet renormalization has been carried out, one analytically continues ε to be *negative*. Then, ε regularizes the IR divergences. IR divergences cancel in IR safe cross sections against corresponding divergences from real emission.

The D-dimensional theory can be thought of as an analytic continuation, starting from the four-dimensional one. There are different ways to define this analytic continuation, which gives rise to different regularization schemes used in the literature. There are also alternatives to dimensional regularization such as Pauli-Villars for ultraviolet divergences, or in some cases to use particle masses to regularize infrared divergences, but we will not use them in these lecture notes.

3.2 Conventions for Minkowski-Space Integrals

Unless otherwise stated, integrals are defined in Minkowski space with 'mostly-minus' metric, i.e. $\eta_{\mu\nu} = \mathrm{diag}(+, -, -, -)$ in four dimensions. Feynman propagators in momentum space are defined as $\sim 1/(p^2 - m^2 + i0) = -1/(-p^2 + m^2 - i0)$, and we will drop the explicit $i0$ from the formulas unless needed for clarification.

When dimensional regularization is used, we set $D = 4 - 2\varepsilon$, with $\varepsilon > 0$ for ultraviolet (UV) and $\varepsilon < 0$ for infrared (IR) divergences. We set the dimensional regularization scale $\mu^2 = 1$. It can always be recovered by dimensional analysis.

At loop level, one encounters integrals over loop momenta, with the integrand being given by propagators and possibly numerator factors. Let us begin by discussing the loop integrals with trivial numerators first.

In order to explain how to define integrals in D-dimensional Minkowski space, we begin with the following example, the tadpole integral

$$\int \frac{d^D k}{i\pi^{D/2}} \frac{1}{(-k^2 + m^2 - i0)^a} = \int \frac{d^{D-1}\mathbf{k}}{i\pi^{D/2}} \int dk_0 \frac{1}{(-k_0^2 + \mathbf{k}^2 + m^2 - i0)^a}, \quad (3.1)$$

where a is arbitrary (for concreteness, one can think of a as a positive integer for now, the formula can then be analytically continued to arbitrary $a \in \mathbb{C}$). Consider the integration over k_0. We see that there are two poles in the complex k_0 plane, at $k_0^\pm = \pm\sqrt{\mathbf{k}^2 + m^2} \mp i0$.[1] We can rotate the contour of integration for k_0 in the complex plane (Wick rotation) without encountering these poles, so that the integration

---

[1]Note that $i0$ is always to be understood as a positive infinitesimal quantity. Strictly speaking the $i0$ in Eq. (3.1) differs from the $i0$ here by a positive factor of $2\sqrt{\mathbf{k}^2 + m^2}$ and terms of order $(i0)^2$ are suppressed. The same will be true when we introduce Feynman parametrization below.

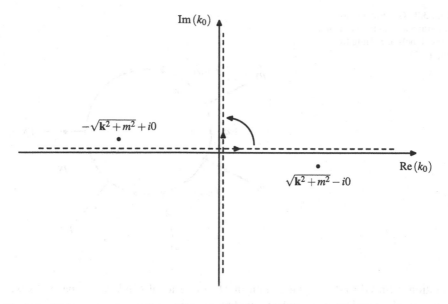

Fig. 3.1 Wick rotation from Minkowski to Euclidean space. We can deform the integration contour from being parallel to the *horizontal axis* to being parallel to the *vertical axis* without encountering the two poles of the propagator indicated by *dots* in the figure

contour becomes parallel to the imaginary axis, $k_0 = ik_{0,E}$, see Fig. 3.1. Defining a Euclidean four-vector $k_E = (k_{0,E}, \mathbf{k})$ we arrive at

$$\int \frac{d^D k_E}{\pi^{D/2}} \frac{1}{(k_E^2 + m^2)^a},$$ (3.2)

where we have dropped the $-i0$ term. For integer D, this integral can be carried out by writing the propagator in the Schwinger parametrization, i.e.

$$\frac{1}{x^a} = \frac{1}{\Gamma(a)} \int_0^\infty d\alpha\, \alpha^{a-1} e^{-\alpha x},$$ (3.3)

and using the known formula for the Gaussian integral in D dimensions. This leads us to

$$\int \frac{d^D k}{i\pi^{D/2}} \frac{1}{(-k^2 + m^2 - i0)^a} = \frac{\Gamma(a - D/2)}{\Gamma(a)} \frac{1}{(m^2 - i0)^{a - D/2}}.$$ (3.4)

Here the Γ function is defined as $\Gamma(z) = \int_0^\infty t^{z-1} e^{-t} dt$. Notice that the dependence on m^2 in Eq. (3.4) is fixed by dimensional analysis. Another simple consistency check can be performed by differentiating w.r.t. m^2, which gives a recursion relation in a. In the following, we will leave the Feynman prescription $-i0$ implicit.

The derivation of Eq. (3.4) assumed integer a and D. We then *define* the integral for non-integer D to be given by analytic continuation. Formula Eq. (3.4) is the key

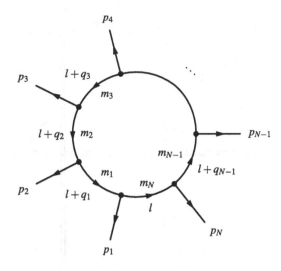

relation we need to define integration in D dimensional Minkowski space. As we
will see, all other integrals can be related to this one.

3.3 General Remarks, Ultraviolet and Infrared Divergences

The Feynman diagrammatic expansion of renormalizable quantum field theories in
four dimensions generically leads to one-loop integrals of the form

$$I_{N,\mathcal{N}}(l)$$
$$= \int \frac{d^D l}{(2\pi)^D} \frac{\mathcal{N}(l^\mu)}{[(l+q_1)^2 - m_1^2 + i0] \cdots [(l+q_{N-1})^2 - m_{N-1}^2 + i0][l^2 - m_N^2 + i0]},$$
$$(3.5)$$

where N counts the number of external lines carrying the outgoing momenta p_i^μ,
compare Fig. 3.2 for the notation. The $N - 1$ region momenta q_i are defined by

$$q_j^\mu = \sum_{k=1}^{j} p_k^\mu, \qquad p_j^\mu = q_j^\mu - q_{j-1}^\mu, \qquad \sum_{k=1}^{N} p_k^\mu = 0 = q_N^\mu. \qquad (3.6)$$

The numerator function $\mathcal{N}(l^\mu)$ is a polynomial in the loop-momentum l^μ and further depends on external data, i.e. the momenta and polarizations of the scattered
particles. Note that the external momenta are in general massive, $p_i^2 = M_i^2$ even
if the external states are massless, e.g. gluons. This happens because an arbitrary
tree-level diagram may be attached to an external $p_i = \sum_{j=j_<}^{j_>} k_j$ line, where the k_i
are the scattered particle's momenta, resulting in a $p_i^2 \neq 0$. This is illustrated in the

following five-gluon triangle graph:

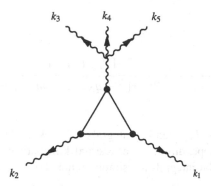

$$p_1 = k_1, \qquad\qquad p_1^2 = 0, \qquad\qquad\qquad (3.7)$$

$$p_2 = k_2, \qquad\qquad p_2^2 = 0, \qquad\qquad\qquad (3.8)$$

$$p_3 = k_3 + k_4 + k_5, \qquad p_3^2 \neq 0. \qquad\qquad (3.9)$$

Also note that fermions in the loop give rise to the same class of integrals Eq. (3.5), as the identity

$$\frac{1}{\not{l} - m} = \frac{\not{l} + m}{l^2 - m^2} \qquad\qquad\qquad (3.10)$$

puts the fermion propagators into the generic form used there.

In a renormalizable four-dimensional quantum field theory the maximal degree of the polynomial $\mathcal{N}(l^\mu)$ in the loop momentum l^μ of an N-point amplitude is N. This is a direct consequence of the Feynman rules in which all vertices are momentum independent—except for the gluon three-point vertex, which is linear in momentum—compare Table 1.2. Hence the maximal power of loop momentum l^μ that may be acquired in the numerator polynomial $\mathcal{N}(l)$ arises from a pure succession of three-gluon vertices, each of which has two legs within the loop and one external leg. The maximal power of l^μ for such a diagram is equal to the number of external legs N. As an example take the following six-gluon one-loop graph:

$$\sim \int \frac{d^4 l}{(2\pi)^4} \frac{\mathcal{O}(l^6)}{l^2 (l - q_1)^2 (l - q_2)^2 \cdots (l - q_5)^2}. \qquad (3.11)$$

If $\mathscr{N}(l)$ in Eq. (3.5) is a constant we speak of a scalar integral. We denote N-point integrals with a uniform polynomial $\mathscr{N}(l)$ of degree r as

$$I_{N,r} = \int \frac{d^D l}{(2\pi)^D}$$

$$\times \frac{\prod_{j=1}^r (u_j \cdot l)}{[(l+q_1)^2 - m_1^2 + i0] \cdots [(l+q_{N-1})^2 - m_{N-1}^2 + i0][l^2 - m_N^2 + i0]}$$

(3.12)

and call the integrals $I_{N,r}$ tensor integrals of rank r. Here the u_i^μ are four-dimensional vectors depending on the external kinematical data of momenta and polarizations. The scalar integrals are simply denoted by $I_N = I_{N,0}$.

3.3.1 Ultraviolet Divergences

Recall that a generic feature of quantum field theory is to have ultraviolet (UV) divergences, i.e. short-distance (in position space) respectively large-momentum divergences. They lead in general to renormalization of bare fields and couplings in the Lagrangian. We refer the reader to standard quantum field theory books for more details.

Let us now see how UV divergences occur in scattering amplitudes. The generic one-loop integral $I_{N,r}$ of Eq. (3.5) will be divergent in the UV, i.e. for large loop momentum l, for high enough rank r. Power counting[2] yields a UV-divergent $I_{N,r}$ when $r \geq 2N - 4$. As $r \leq N$ from the discussion above we deduce that UV divergences appear for $N \geq 2N - 4$ or $N \leq 4$. Hence at one loop, only one-, two-, three- or four-point integrals can be divergent in the UV, *all* five and higher-point integrals are UV finite. If we restrict to scalar integrals we see that only the one- and two-point scalar integrals are UV divergent.

3.3.2 Infrared Divergences

As was already mentioned in the introduction, in on-shell processes involving massless particles there can also be infrared divergences. They occur in regions of the loop integration when a sufficient number of propagators in the integrand go on

---

[2]For this counting one simply adds the powers of the loop-momentum l in the denominator and numerator of the integral including the measure. A positive or vanishing result indicates a UV-divergent integral: $\int d^4 l\, l^a / l^b$ is UV-divergent for $4 + a \geq b$. Equality in this relation entails a logarithmic divergence, the other cases yield power-like divergences. In fact dimensional regularization is only sensitive to logarithmic divergencies, power-like divergences are set to zero.

mass shell simultaneously, thereby introducing a singularity that is not balanced by the measure factor.

Let us give two examples. The first example is a massive one-loop form factor integral

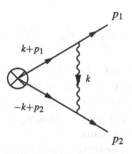

$$\int \frac{d^D k}{i\pi^{D/2}} \frac{1}{k^2(-(k+p_1)^2+m^2)(-(k-p_2)^2+m^2)}$$

$$= \int \frac{d^D k}{i\pi^{D/2}} \frac{1}{k^2(-k^2-2k\cdot p_1)(-k^2+2k\cdot p_2)}, \tag{3.13}$$

where we have used the on-shell conditions $p_i^2 = m^2$. This integral could occur for example in a QED form factor of massive electrons, where the massless propagator $1/k^2$ corresponds to a photon exchange. We can see that when all components of the loop momentum k^μ become small, the integral develops a logarithmic divergence for $D = 4$. In $D = 4 - 2\varepsilon$, with $\varepsilon < 0$, this divergence is regularized and produces a single pole, i.e. $1/\varepsilon$. This divergence, and the region of integration associated to it is called 'soft'. The limit $k^\mu \to 0$ is the 'soft' limit.

The second example is the same integral in the massless case $m = 0$. In this case there are additional, collinear divergences coming from regions of integrations where the loop momentum k^μ becomes collinear to one of the external null momenta p_i^μ: let $k^\mu = c p_1^\mu$ then the denominator in Eq. (3.13) develops two quadratic poles, $k^2 = c^2 p_1^2 = 0$ and $(k + p_1)^2 = (1 + c)^2 p_1^2 = 0$ as $p_1^2 = 0$. Since the soft divergence for small k^μ is still present, this leads to a double pole in ε.

Infrared divergences of scattering amplitudes are closely related to ultraviolet properties of certain Wilson loops [1]. This will be discussed in the Chap. 4.

3.3.3 Regularization Scheme

There are different regularization schemes for analytically continuing the four-dimensional theory to D dimensions. In the four-dimensional helicity scheme, all calculations are done in D dimensions, except that the external states are kept in four dimensions, as is the Dirac matrix algebra. The reason behind this is that in this way one can continue to use the four-dimensional spinor helicity techniques that have proven very useful at tree-level. (A discussion of different regularization schemes is given in Ref. [2].) We shall adopt this scheme in the sequel.

In particular, the loop integration measure in the Feynman rules then becomes[3]

$$\int \frac{d^4 l}{(2\pi)^4} \longrightarrow \int \frac{d^D l}{(2\pi)^D} \quad \text{with } D = 4 - 2\varepsilon. \tag{3.14}$$

A word of caution is in order here. It should go without saying that, in order to obtain a consistent answer, the regularization has to be introduced at the very beginning of the calculation. In particular, we stress that doing the numerator algebra naively in four dimensions, and then just replacing the loop integration measure as in Eq. (3.14) will in general miss terms.

The one-loop four-gluon all positive helicity amplitude $A_4(1^+, 2^+, 3^+, 4^+)$ in pure Yang-Millls theory is a case in point. In this regularization scheme, it can be seen to be non-zero precisely due to the mismatch in dimensionality between external and internal momenta.

3.4 Integral Reduction

We now turn to the reduction of generic one-loop tensor integrals occurring in renormalizable four dimensional quantum field theories to a set of basis integrals. As we will see, if one is interested only in obtaining a result for scattering amplitudes up to $\mathcal{O}(\varepsilon)$ terms, one can reduce a generic one-loop integral $I_{N,\mathcal{N}(l)}$ to a linear combination of one-loop *scalar* integrals of four-, three-, two- and one-point type, and terms that are rational in the external variables. This rational part we denote by $\mathcal{R}$. This result goes back to Passarino and Veltman [3] and states the decomposition

$$I_{N,\mathcal{N}(l)} = \sum_{j_4} c_{4;j_4} I_4^{(j_4)} + \sum_{j_3} c_{3;j_3} I_3^{(j_3)} + \sum_{j_2} c_{2;j_2} I_2^{(j_2)}$$
$$+ \sum_{j_1} c_{1;j} I_1^{(j_1)} + \mathcal{R} + \mathcal{O}(\varepsilon), \tag{3.15}$$

where $I_N^{(j)}$ denotes the scalar N-point integral of type 'j' which refers to the various possible distributions of the external momenta p_i on the N-legs of $I_N^{(j)}$. The coefficients $c_{N,j}$ are algebraic four-dimensional quantities related to tree-level amplitudes. Equation (3.15) is the central result of the one-loop integral reduction to box, triangle, bubble and tadpole integrals. We shall derive it in the following.

The one-loop basis of scalar integrals appearing in this decomposition reads:

---

[3]The definition will be clarified below in the Feynman representation.

- Tadpoles:

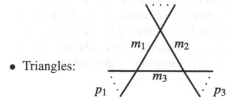

$$m_1$$

$$I_1(m_1^2) = \int \frac{d^D l}{(2\pi)^D} \frac{1}{d_1}. \qquad (3.16)$$

- Bubbles:

$$m_1$$
$$p_1$$
$$m_2$$

$$I_2(p_1^2; m_1^2, m_2^2) = \int \frac{d^D l}{(2\pi)^D} \frac{1}{d_1 d_2}. \qquad (3.17)$$

- Triangles:

$$p_2$$
$$m_1 \quad m_2$$
$$m_3$$
$$p_1 \qquad p_3$$

$$I_3(p_1^2, p_2^2, p_3^2; m_1^2, m_2^2, m_3^2) = \int \frac{d^D l}{(2\pi)^D} \frac{1}{d_1 d_2 d_3}. \qquad (3.18)$$

- Boxes:

$$p_2 \qquad p_3$$
$$m_2$$
$$m_1 \qquad m_3$$
$$m_4$$
$$p_1 \qquad p_4$$

$$I_4(p_1^2, p_2^2, p_3^2, p_4^2; s_{12}, s_{23}; m_1^2, m_2^2, m_3^2, m_4^2) = \int \frac{d^D l}{(2\pi)^D} \frac{1}{d_1 d_2 d_3 d_4}, \qquad (3.19)$$

where

$$d_i = (l + q_i)^2 - m_i^2 + i0, \qquad s_{ij} = (p_i + p_j)^2, \qquad q_n = \sum_{i=1}^{n} p_i, \qquad q_N = 0.$$

All these dimensionally regulated integrals are known in analytic form and may be found e.g. in Ref. [4]. There also exist a number of numerical implementations of these integrals to be found in [5–8]. We note that UV divergences only occur in the tadpole and bubble-type integrals, IR divergences can only occur in triangles and boxes.

3.4.1 The Van Neerven-Vermaseren Basis

The next goal is to show why the integral reduction formula Eq. (3.15) holds true. This can be seen in a two step procedure. In the first step we want to reduce N-point tensor-integrals of higher-rank r to scalar integrals with N points and lower:

$$\int \frac{d^D l}{(2\pi)^D} \frac{l^{\mu_1} l^{\mu_2} \cdots l^{\mu_r}}{d_1 \cdots d_N} \longrightarrow \sum_{N' \leq N} \int \frac{d^D l}{(2\pi)^D} \frac{1}{d_1 \cdots d_{N'}}. \tag{3.20}$$

The second step is then the reduction of scalar integrals with $N > 4$ points to those with $N = 1, 2, 3, 4$ as Eq. (3.15) states.

The key idea in the tensor integral reduction is to expand the loop momentum l^μ in a convenient basis determined by the external region momenta q_i. As there are $N - 1$ linear independent q_i in an N-point integral the q_i alone will not be enough to specify the basis completely as long as $N \leq 4$. Equivalently one could also choose the inflowing momenta p_i for spanning a basis, however, the region momenta turn out to be the more convenient choice.

3.4.1.1 Two Dimensional Example

For pedagogical purposes let us first discuss the simpler two-dimensional case. Here, from three-point functions onwards there are two linearly independent region momenta q_1 and q_2 in the problem. Now in principle we could decompose $l^\mu = c_1 q_1^\mu + c_2 q_2^\mu$, but as the q_i do not span an orthonormal basis ($q_i \cdot q_j \neq \delta_{ij}$), the coefficients c_i are not easily determined. A clever basis choice leading to a simple coefficient structure is to employ the Levi-Civita tensor as follows: Consider Schouten's identity in the form

$$l^\mu \varepsilon^{\nu_1 \nu_2} = l^{\nu_1} \varepsilon^{\mu \nu_2} + l^{\nu_2} \varepsilon^{\nu_1 \mu}, \tag{3.21}$$

contracting this relation with the two region momenta $q_{1\nu_1}$ and $q_{2\nu_2}$ leads to

$$l^\mu = (l \cdot q_1) \frac{\varepsilon^{\mu q_2}}{\varepsilon^{q_1 q_2}} + (l \cdot q_2) \frac{\varepsilon^{q_1 \mu}}{\varepsilon^{q_1 q_2}}. \tag{3.22}$$

Note that in the above we are using a convenient short-hand notation $\varepsilon^{\mu \nu_1} q_{2\nu_1} = \varepsilon^{\mu q_2}$ 'soaking' up the contracted indices by pulling the four-vectors into the tensor index positions. We hence see that using the dual basis vectors

$$v_1^\mu = \frac{\varepsilon^{\mu q_2}}{\varepsilon^{q_1 q_2}}, \qquad v_2^\mu = \frac{\varepsilon^{q_1 \mu}}{\varepsilon^{q_1 q_2}}, \tag{3.23}$$

leads to the convenient decomposition $l^\mu = \sum_{i=1}^{2}(q_i \cdot l)v_i^\mu$. The dual basis v_j satisfies the relations $(q_i \cdot v_j) = \delta_{ij}$ but $(q_i \cdot q_j) \neq \delta_{ij} \neq (v_i \cdot v_j)$.

This expansion is very useful as the scalar products $(l \cdot q_i)$ may be expressed as linear combinations of inverse propagators and external scalars

$$l \cdot q_i = \frac{1}{2}\left[\left((l+q_i)^2 - m_i^2 + i0\right) - \left(l^2 - m_N^2 + i0\right) - \left(q_i^2 - m_i^2\right) - m_N^2\right] \quad (3.24)$$

$$= \frac{1}{2}\left[d_i - d_N - q_i^2 + m_i^2 - m_N^2\right]. \quad (3.25)$$

This reduces any tensor integral with more than two points to a scalar integral by decomposing the loop momentum l^μ appearing in the numerator of the tensor integrals as

$$l^\mu = \frac{1}{2}\sum_{i=1}^{2}\left[d_i - d_N - q_i^2 + m_i^2 - m_N^2\right]v_i^\mu. \quad (3.26)$$

There is one important caveat in this argument: the vector l^μ is in fact defined in $D = 2 - 2\varepsilon$ dimensions. But the construction of the basis vector v_i in Eq. (3.23) only exists in $D = 2$. This is obvious through the appearance of Levi-Civita-tensors. In order to generalize this construction we write Eq. (3.23) in a dimension independent fashion. Starting with

$$v_1^\mu = \frac{\varepsilon_{q_1 q_2}\varepsilon^{\mu q_2}}{\varepsilon_{q_1 q_2}\varepsilon^{q_1 q_2}}, \qquad v_2^\mu = \frac{\varepsilon_{q_1 q_2}\varepsilon^{q_1 \mu}}{\varepsilon_{q_1 q_2}\varepsilon^{q_1 q_2}}, \quad (3.27)$$

one uses the identity

$$\varepsilon^{\mu_1\mu_2}\varepsilon_{\nu_1\nu_2} = \delta_{\nu_1}^{\mu_1}\delta_{\nu_2}^{\mu_2} - \delta_{\nu_2}^{\mu_1}\delta_{\nu_1}^{\mu_2} = \det\left(\delta_\nu^\mu\right) =: \delta_{\nu_1\nu_2}^{\mu_1\mu_2}, \quad (3.28)$$

introducing the generalized Kronecker symbol $\delta_{\nu_1\nu_2}^{\mu_1\mu_2}$ which is antisymmetric in its upper and lower indices. Hence we can rewrite the two-dimensional basis vectors of Eq. (3.23) as

$$v_1^\mu = \frac{\delta_{q_1 q_2}^{\mu q_2}}{\Delta_2}, \qquad v_2^\mu = \frac{\delta_{q_1 q_2}^{q_1 \mu}}{\Delta_2}, \quad \text{with } \Delta_2 := \delta_{q_1 q_2}^{q_1 q_2} = q_1^2 q_2^2 - (q_1 \cdot q_2)^2, \quad (3.29)$$

embedding the dual basis vectors in a $(D = 2 + d_t)$-dimensional space with $d_t = -2\varepsilon$ formally.

3.4.1.2 The General Case

We now generalize the construction of the van Neerven-Vermaseren basis to the $D = d_p + d_t$ dimensional case. Here d_p denotes the number of physical dimensions and d_t is the number of transverse dimensions. It is important to realize that the value of d_p depends on the number of points N of the integral in question. One has

$d_p = \min(N - 1, 4)$ as there are $(N - 1)$ q_i's in an N point integral and one can have at most four linearly independent q_i's in four dimensions. Recall that in our scheme the external momenta remain in four dimensions.

We begin with writing down the van Neerven-Vermaseren basis of the d_p dimensional subspace spanned by the dual vectors to $q_1, q_2, \ldots, q_{d_p}$

$$v_i^\mu(q_1, \ldots, q_{d_p}) := \frac{\delta_{q_1 \cdots q_{i-1} q_i q_{i+1} \cdots q_{d_p}}^{q_1 \cdots q_{i-1} \mu q_{i+1} \cdots q_{d_p}}}{\Delta_{d_p}(q_1, \ldots, q_{d_p})}, \quad i = 1, \ldots, d_p, \tag{3.30}$$

where we have defined $\Delta_{d_p}(q_1, \ldots, q_{d_p}) = \det(q_i \cdot q_j)$ along with the generalized Kronecker symbol

$$\delta_{\nu_1 \cdots \nu_{d_p}}^{\mu_1 \cdots \mu_{d_p}} = \begin{vmatrix} \delta_{\nu_1}^{\mu_1} & \delta_{\nu_2}^{\mu_1} & \cdots & \delta_{\nu_{d_p}}^{\mu_1} \\ \delta_{\nu_1}^{\mu_2} & \delta_{\nu_2}^{\mu_2} & \cdots & \delta_{\nu_{d_p}}^{\mu_2} \\ \vdots & \vdots & & \vdots \\ \delta_{\nu_1}^{\mu_{d_p}} & \delta_{\nu_2}^{\mu_{d_p}} & \cdots & \delta_{\nu_{d_p}}^{\mu_{d_p}} \end{vmatrix}, \tag{3.31}$$

which is completely antisymmetric in the upper and lower indices. This is a dual basis obeying

$$v_i \cdot q_j = \delta_{ij}, \quad \text{for } i, j \in \{1, \ldots, d_p\}. \tag{3.32}$$

The remaining transverse space is Euclidean and spanned by a set of d_t basis vectors which we denote by n_r^μ, $r = 1, \ldots, d_t$, and require to obey the orthonormality conditions

$$n_r \cdot n_s = \delta_{rs}, \quad v_i \cdot n_r = 0 \quad \text{and} \quad q_i \cdot n_r = 0. \tag{3.33}$$

Summarizing, we have the general dual basis expansion for the $D = d_p + d_t$ loop momentum

$$l^\mu = \sum_{i=1}^{d_p} (l \cdot q_i) v_i^\mu + \sum_{r=1}^{d_t} (l \cdot n_i) n_i^\mu, \tag{3.34}$$

and we note once more the important relation $(l \cdot q_i) = \frac{1}{2}[d_i - d_N - q_i^2 + m_i^2 - m_N^2]$.

3.4.2 Reduction of the Integrand in $D = 4 - 2\varepsilon$ Dimensions

A generic N point tensor integral of rank r has the integrand, cf. Eq. (3.12)

$$\frac{\prod_{i=1}^r u_i \cdot l}{d_1 d_2 \cdots d_N}, \quad r \leq N. \tag{3.35}$$

Here the r u_i^μ are vectors made of the external momenta and polarizations. In particular they lie in a four dimensional subspace of the D-dimensional regulating space of l^μ.

3.4.2.1 Tensor Integrals with $N \geq 5$ Points

For more than four points we can use the first four linearly independent region momenta q_1, q_2, q_3, q_4 to span the van Neerven-Vermaseren basis v_i^μ with $d_p = 4$

$$l^\mu = \sum_{i=1}^{4}(l \cdot q_i)v_i^\mu + (l \cdot n_\varepsilon)n_\varepsilon^\mu. \tag{3.36}$$

Here n_ε^μ spans the unphysical regulating $d_t = -2\varepsilon$ dimensional subspace. Due to $u_i \cdot n_\varepsilon = 0$, since the u_i depend on external data and thus have no dependence on the unphysical dimensions, we have for every factor appearing in the denominator of Eq. (3.35)

$$u_i \cdot l = \sum_{j=1}^{4}(l \cdot q_j)(v_j \cdot u_i)$$

$$= \frac{1}{2}\sum_{j=1}^{4}[d_j - q_j^2 + m_j^2](v_j \cdot u_i) - \frac{1}{2}(d_N + m_N^2)\sum_{j=1}^{4}(v_j \cdot u_i). \tag{3.37}$$

As the u_i and v_j are independent of the loop momentum l^μ we see that the integrand Eq. (3.35) with $N \geq 5$ can be completely reduced to a sum of scalar integrands with $N' \leq N$ points by inserting the above identity. We have thus shown that tensor integrals with $N > 4$ points can be reduced to *scalar* integrals with up to N points.

3.4.2.2 Scalar Integrals with $N > 5$ Points

Let us start out with a scalar integrand of the form

$$I_N = \prod_{i=1}^{N}\frac{1}{d_i}, \quad N > 5, \ d_i = (l + q_i)^2 - m_i^2 + i0. \tag{3.38}$$

The goal is to reduce this integrand to a sum of lower-point scalar integrands with maximally five points. For this consider the following five equations for N unknowns α_i

$$\sum_{i=1}^{N}\alpha_i = 0, \quad \sum_{i=1}^{N}\alpha_i q_i^\mu = 0. \tag{3.39}$$

As $N > 5$ we have at least one non-trivial solution of these equations for the set of N α_i's.[4] Picking one such solution we have

$$\sum_{i=1}^{N} \alpha_i d_i = \sum_{i=1}^{N} \alpha_i \left(l^2 + 2l \cdot q_i + q_i^2 - m_i^2 + i0 \right) = \sum_{i=1}^{N} \alpha_i \left(q_i^2 - m_i^2 + i0 \right), \quad (3.40)$$

or

$$1 = \frac{\sum_{i=1}^{N} \alpha_i d_i}{\sum_{i=1}^{N} \alpha_i \left(q_i^2 - m_i^2 + i0 \right)}. \quad (3.41)$$

Note that in this expression the α_i and the denominator are independent of the loop momentum l^μ. Inserting this 1 into the integrand of Eq. (3.38) yields

$$I_N \cdot 1 = \sum_{i=1}^{N} \left(\prod_{j \neq i} \frac{1}{d_j} \right) \frac{\alpha_i}{\sum_{i=1}^{N} \alpha_i \left(q_i^2 - m_i^2 + i0 \right)} = \sum_{k=1}^{N} c_k I_{N-1,k}, \quad (3.42)$$

where $I_{N-1;k} = \prod_{j=1, j \neq k}^{N} \frac{1}{d_j}$ is a scalar integral with $N - 1$ points. We have thus reduced the $N > 5$ point scalar integrand to a sum of scalar integrands with $N - 1$ points. Iterating this procedure it is clear that one may reduce any scalar integrand with $N > 5$ to a sum of scalar pentagons only.

Summarizing, what we have shown so far is that any one-loop integrand of a renormalizable quantum field theory in $D = 4 - 2\varepsilon$ dimensions may be reduced to scalar pentagons and tensor integrals of up to four points, i.e.

$$I_{N \geq 5, r}^{\text{any}} = c_5 I_5^{\text{scalar}} + \sum_{N=1}^{4} \sum_{r'=0}^{N} c_{N,r'} I_{N,r'}^{\text{tensor}}. \quad (3.43)$$

What remains to be done is to show that the tensor integrals with $N \leq 4$ points may be reduced to scalar integrals of no higher number of points. Finally one will have to show that the scalar pentagon may be expressed in terms of scalar boxes, triangles and bubbles too.

3.4.2.3 Tensor Integrals with $N = 4$ Points

We now consider the reduction of a 4-point tensor integral of rank $r \leq 4$

$$I_{4,r} = \int \frac{d^D l}{(2\pi)^D} \frac{\mathcal{N}_{4,r}(l)}{d_1 d_2 d_3 d_4}, \qquad \mathcal{N}_{4,r}(l) = \prod_{i=1}^{r} u_i \cdot l, \quad (3.44)$$

---

[4] In fact $q_N^\mu = 0$ so in the last equation the sum effectively only runs up to $i = N - 1$, but this does not change the argument.

where the u_i only depend on external four dimensional data. The physical space of this integral is $d_p = 3$ dimensional (spanned by the region momenta q_1, q_2 and q_3) while the transverse space is $d_t = 1 - 2\varepsilon$ dimensional, hence

$$l^\mu = \sum_{i=1}^{3} (l \cdot q_i) v_i^\mu + (l \cdot n_4) n_4^\mu + (l \cdot n_\varepsilon) n_\varepsilon^\mu. \tag{3.45}$$

As $u_i \cdot n_\varepsilon = 0$ we have

$$u_i \cdot l = \sum_{j=1}^{3} (u_i \cdot v_j)(l \cdot q_j) + (u_i \cdot n_4)(l \cdot n_4). \tag{3.46}$$

Inserting the relation $l \cdot q_j = \frac{1}{2}(d_j - d_N) + \text{const}$, where "const" refers to l independent terms, into the first term in the above yields $u_i \cdot l = (u_i \cdot n_4)(l \cdot n_4) + \mathcal{O}(d_i) + \text{const}$, where $\mathcal{O}(d_i)$ refers to terms proportional to an inverse propagator which will lower the point number of the integrand. Inserting this into $\mathcal{N}_{4,r}(l)$ and keeping only the 4-point integrands we have

$$\frac{\mathcal{N}_{4,r}(l)}{d_1 d_2 d_3 d_4} = \frac{\prod_{i=1}^{r} (u_i \cdot l)}{d_1 d_2 d_3 d_4} = \sum_{i=1}^{r} c_i \frac{(l \cdot n_4)^i}{d_1 d_2 d_3 d_4} + \frac{\text{const}}{d_1 d_2 d_3 d_4} + \begin{pmatrix} \text{integrands of} \\ \text{lower points} \end{pmatrix}. \tag{3.47}$$

Here the last term contains at least one denominator d_i leading to lower-point integrands, which will be reduced later. In order to simplify this further we now square Eq. (3.45) and use $v_i \cdot n_j = 0$, $n_4 \cdot n_\varepsilon = 0$ and $l^2 = d_N + m_N^2$ as well as $l \cdot q_i = \frac{1}{2}(d_i - d_N) + \text{const}$ to find the useful relation

$$(l \cdot n_4)^2 = -(l \cdot n_\varepsilon)^2 + \text{const} + \mathcal{O}(d_i). \tag{3.48}$$

Hence in Eq. (3.47) the appearance of $(l \cdot n_4)^2$ may be traded off by $(l \cdot n_\varepsilon)^2$ and we may write

$$\mathcal{N}_{4,r\leq 4}(l) = \tilde{d}_0 + \tilde{d}_1 (l \cdot n_4) + \tilde{d}_2 (l \cdot n_\varepsilon)^2 + \tilde{d}_3 (l \cdot n_4)(l \cdot n_\varepsilon)^2$$

$$+ \tilde{d}_4 (l \cdot n_\varepsilon)^4 + \begin{pmatrix} \text{integrands of} \\ \text{lower points} \end{pmatrix}, \tag{3.49}$$

where the $\tilde{d}_i$ are coefficients depending on the external data. At this point it is not entirely clear what we have gained by inserting Eq. (3.48) into $\mathcal{N}_{r\leq 4}(l)$. We will later show, however, that all remaining l-dependent terms in the above integrate to zero or contribute to the rational part $\mathcal{R}$ in Eq. (3.15).

3.4.2.4 Tensor Integrals with $N = 3$ Points

Similar considerations for the 3-point tensor-integrals of rank $r \leq 3$ using its $d_p = 2$ dimensional physical space with dual basis v_1^μ and v_2^μ result in an expansion of the loop momentum

$$l^\mu = \sum_{i=1}^{2}(l \cdot q_i)v_i^\mu + (l \cdot n_3)n_3^\mu + (l \cdot n_4)n_4^\mu + (l \cdot n_\varepsilon)n_\varepsilon^\mu. \tag{3.50}$$

In analogy to the steps above for the $N = 4$ case and using the relation

$$(l \cdot n_3)^2 + (l \cdot n_4)^2 = -(l \cdot n_\varepsilon)^2 + \text{const} + \text{lower-point integrands} \tag{3.51}$$

allows one to rewrite the numerator factor as

$$\mathcal{N}_{3,r\leq3}(l) = \tilde{c}_0 + \tilde{c}_1(l \cdot n_3) + \tilde{c}_2(l \cdot n_4) + \tilde{c}_3(l \cdot n_\varepsilon)^2 + \tilde{c}_4\big[(l \cdot n_3)^2 - (l \cdot n_4)^2\big]$$
$$+ \tilde{c}_5(l \cdot n_3)(l \cdot n_4) + \tilde{c}_6(l \cdot n_3)^3 + \tilde{c}_7(l \cdot n_4)^3 + \tilde{c}_8(l \cdot n_4)^2(l \cdot n_\varepsilon)$$
$$+ \tilde{c}_9(l \cdot n_3)^2(l \cdot n_\varepsilon) + \text{lower-point integrands}, \tag{3.52}$$

with coefficients $\tilde{c}_i$ depending on the external data. Similar results hold for the remaining two-point and one-point tensor integrals, see [9] for a derivation.

3.4.2.5 Lower-Point Tensor Integrals

Without derivation we simply quote the reduction of the 2-point tensor integrals of rank $r \leq 2$ in terms of the numerator

$$\mathcal{N}_{2,r\leq2}(l) = \tilde{b}_0 + \tilde{b}_1(l \cdot n_2) + \tilde{b}_2(l \cdot n_3) + \tilde{b}_3(l \cdot n_4) + \tilde{b}_4(l \cdot n_\varepsilon)^2$$
$$+ \tilde{b}_5\big[(l \cdot n_2)^2 - (l \cdot n_4)^2\big] + \tilde{b}_6\big[(l \cdot n_3)^2 - (l \cdot n_4)^2\big]$$
$$+ \tilde{b}_7(l \cdot n_2)(l \cdot n_3) + \tilde{b}_8(l \cdot n_2)(l \cdot n_4) + \tilde{b}_9(l \cdot n_3)(l \cdot n_4). \tag{3.53}$$

For the one-point tensor integral there is no real reduction possible as the number of physical dimensions vanish, thus one has

$$\mathcal{N}_{1,r\leq1}(l) = \tilde{a}_0 + \tilde{a}_1(l \cdot n_1) + \tilde{a}_2(l \cdot n_2) + \tilde{a}_3(l \cdot n_3) + \tilde{a}_4(l \cdot n_4), \tag{3.54}$$

where it is clear that only the $\tilde{a}_0$ coefficient is relevant as all other terms integrate to zero.

3.4.2.6 Reduction of the Five-Point Scalar Integral to Lower Point Integrals

What remains to be shown is why we do not need to include the 5-point (pentagon) scalar-integral into our set of master integrals. We have

$$I_5 = \int \frac{d^D l}{(2\pi)^D} \frac{1}{d_1 d_2 d_3 d_4 d_5}. \tag{3.55}$$

The physical dimension of this integral is four hence the loop momentum l^μ may be expanded as

$$l^\mu = \sum_{i=1}^{4} v_i^\mu (l \cdot q_i) + (l \cdot n_\varepsilon) n_\varepsilon^\mu = \frac{1}{2} \sum_{i=1}^{4} v_i^\mu \left(d_i - d_5 + m_i^2 - m_5^2 - q_i^2\right) + (l \cdot n_\varepsilon) n_\varepsilon^\mu.$$
(3.56)

Squaring this relation yields with $l^2 = d_5 + m_5$ the useful representation

$$(l \cdot n_\varepsilon)^2 = d_5 + m_5$$

$$- \frac{1}{4} \sum_{i,j=1}^{4} (v_i \cdot v_j)\left(d_i - d_5 + m_i^2 - m_5^2 - q_i^2\right)\left(d_j - d_5 + m_j^2 - m_5^2 - q_j^2\right),$$
(3.57)

which tells us in turn that

$$(l \cdot n_\varepsilon)^2 = \text{const} + \mathcal{O}(\{d_i\}) + \mathcal{O}(\{d_i d_j\}),$$
(3.58)

where "const" as usual denotes l^μ independent terms. Note that both the $\mathcal{O}(\{d_i\})$ and the $\mathcal{O}(\{d_i d_j\})$ terms will cancel at least one propagator reducing the five point integrand to a lower number. Special attention is needed for the case when one has the same inverse propagator squared, i.e. a term d_i^2 appearing in $\mathcal{O}(\{d_i d_j\})$. Then one may rewrite the resulting integral in terms of a four-point tensor integral, which in turn is reducible to scalar integrands with at most 4 points. Inserting the 1 arising from Eq. (3.58)

$$1 = \frac{(l \cdot n_\varepsilon)^2}{\text{const}} + \mathcal{O}(\{d_i\}) + \mathcal{O}(\{d_i d_j\})$$
(3.59)

into Eq. (3.55) teaches us that

$$I_5 = \frac{1}{\text{const}} \int \frac{d^D l}{(2\pi)^D} \frac{(l \cdot n_\varepsilon)^2}{d_1 d_2 d_3 d_4 d_5} + \text{lower-point scalar integrals.}$$
(3.60)

The remaining five-point integral on the right-hand-side may be shown to vanish in the $D \to 4$ limit, as we will discuss in the next section. Anticipating this result we see that it is possible to express the pentagon integral I_5 in terms of lower-point integrals, up to terms of $\mathcal{O}(\varepsilon)$. Therefore, it is not required in the four-dimensional integral basis. If one is interested in finite ε contributions, however, one needs to keep the scalar pentagon integral in the basis.

3.4.3   Higher Dimensional Loop Momentum Integration

We now want to show that the higher dimensional loop momentum integrations of the obtained reduced forms of the integrands Eqs. (3.49), (3.52), (3.53) and (3.54) lead to the promised representation in terms of scalar integrals. For this we consider the angular integrations in the transverse space $\mathbb{R}^{d_t}$.

To begin with consider the 4-point case resulting from Eq. (3.49)

$$I_4^{\text{tensor}} = \int \frac{d^{d_p+d_t}l}{(2\pi)^{d_p+d_t}} \frac{1}{d_1 d_2 d_3 d_4} \Big[\tilde{d}_0 + \tilde{d}_1 (l \cdot n_4) + \tilde{d}_2 (l \cdot n_\varepsilon)^2$$
$$+ \tilde{d}_3 (l \cdot n_4)(l \cdot n_\varepsilon)^2 + \tilde{d}_4 (l \cdot n_\varepsilon)^4 \Big]. \tag{3.61}$$

Here $d_p = 3$ denotes the number of physical dimensions of the integral which are spanned by v_i^μ (or the q_i^μ) and $d_t = 1 - 2\varepsilon$ is the number of transverse dimensions. We have

$$d_i = (l + q_i)^2 - m_i^2 = l_\perp^2 + (l_\parallel + q_1)^2 - m_i^2, \tag{3.62}$$

where

$$l_\perp^\mu = n_4^\mu (l \cdot n_4) + n_\varepsilon^\mu (l \cdot n_\varepsilon) \in \mathbb{R}^{d_t} \quad \text{and} \quad l_\parallel^\mu = \sum_{i=1}^3 v_i^\mu (l \cdot q_i) \in \mathbb{R}^{d_p-1,1}, \tag{3.63}$$

as $q_i \cdot n_4 = 0 = q_i \cdot n_\varepsilon$. The numerator in the square brackets of Eq. (3.61) depends only on the transverse directions of the loop momentum $\mathscr{N}_{r \le 4}(l) = \mathscr{N}_{r \le 4}(l_\perp)$, which is a general property holding for all reduced lower-point tensor integrals. At the same time the inverse propagators appearing in the integrand of Eq. (3.61) depend on the transverse directions only via the scalar quantity $l_\perp^2$. This simple observation entails an important simplification of the integral which is of the generic form

$$I_n^{\text{tensor}} = \int \frac{d^{d_p}l_\parallel d^{d_t}l_\perp}{(2\pi)^D} P(l_\parallel^\mu, l_\perp^2) \mathscr{N}_r(l_\perp^\mu), \quad \text{with } P(l_\parallel^\mu, l_\perp^2) = \frac{1}{d_1 \cdots d_n}. \tag{3.64}$$

By rotational symmetry in the transverse space we have

$$\int \frac{d^{d_t}l}{(2\pi)^{d_t}} P(l_\parallel^\mu, l_\perp^2) \begin{pmatrix} l_\perp^\mu \\ l_\perp^{\mu_1} l_\perp^{\mu_2} \\ l_\perp^{\mu_1} l_\perp^{\mu_2} l_\perp^{\mu_3} \\ l_\perp^{\mu_1} l_\perp^{\mu_2} l_\perp^{\mu_3} l_\perp^{\mu_4} \end{pmatrix}$$

$$= \int \frac{d^{d_t}l}{(2\pi)^{d_t}} P(l_\parallel^\mu, l_\perp^2) \cdot \begin{pmatrix} 0 \\ c_1 \eta_\perp^{\mu_1 \mu_2} l_\perp^2 \\ 0 \\ c_2 (\eta_\perp^{\mu_1 \mu_2} \eta_\perp^{\mu_3 \mu_4} + \eta_\perp^{\mu_1 \mu_3} \eta_\perp^{\mu_2 \mu_4} + \eta_\perp^{\mu_1 \mu_4} \eta_\perp^{\mu_2 \mu_3})(l_\perp^2)^2 \end{pmatrix},$$
$$\tag{3.65}$$

with $c_1 = d_t^{-1}$ and $c_2 = (d_t^2 + 2d_t)^{-1}$. Applying this insight to the 4-point reduced tensor-integral Eq. (3.61) shows that the $\tilde{d}_1$ and $\tilde{d}_3$ contributions cancel and we are left with

$$I_4^{\text{tensor}} = \tilde{d}_0 I_4 + \tilde{d}_2 \int \frac{d^D l}{(2\pi)^D} \frac{(l \cdot n_\varepsilon)^2}{d_1 d_2 d_3 d_4} + \tilde{d}_4 \int \frac{d^D l}{(2\pi)^D} \frac{(l \cdot n_\varepsilon)^4}{d_1 d_2 d_3 d_4}, \tag{3.66}$$

where I_4 denotes the scalar box integral of Eq. (3.19). The remaining two integrals contribute to the rational part $\mathscr{R}$ of the amplitude. Indeed as $n_\varepsilon \cdot q_i = 0$ both integrals have to be proportional to n_ε^2 and $(n_\varepsilon^2)^2$ respectively. As formally $n_\varepsilon^2 = -2\varepsilon$ we see that in the $\varepsilon \to 0$ limit these terms can only contribute if the integral diverges. By power counting only the second integral in the above is divergent, hence we conclude

$$I_4^{\text{tensor}} = \tilde{d}_0 I_4 + \tilde{d}_4 \int \frac{d^D l}{(2\pi)^D} \frac{(l \cdot n_\varepsilon)^4}{d_1 d_2 d_3 d_4} + \mathscr{O}(\varepsilon), \tag{3.67}$$

and the contribution to $\mathscr{R}$ is proportional to $\tilde{d}_4$.

Analogous considerations for the 3-point case yield

$$I_3^{\text{tensor}} = \tilde{c}_0 I_3 + \tilde{c}_3 \int \frac{d^D l}{(2\pi)^D} \frac{(l \cdot n_\varepsilon)^2}{d_1 d_2 d_3} + \mathscr{O}(\varepsilon), \tag{3.68}$$

with the triangle integral I_3 of Eq. (3.18), while for the 2-point case one finds the compact result

$$I_2^{\text{tensor}} = \tilde{b}_0 I_2 + \tilde{b}_4 \int \frac{d^D l}{(2\pi)^D} \frac{(l \cdot n_\varepsilon)^2}{d_1 d_2} + \mathscr{O}(\varepsilon), \tag{3.69}$$

with the bubble integral I_2 of Eq. (3.17).

After this discussion we may now return to the open issue of Sect. 3.4.2.6 showing that

$$\lim_{\varepsilon \to 0} \int \frac{d^D l}{(2\pi)^D} \frac{(l \cdot n_\varepsilon)^2}{d_1 d_2 d_3 d_4 d_5} = 0. \tag{3.70}$$

Again this is seen by power counting: the integral is UV finite near $D = 4$ and has to be proportional to $n_\varepsilon^2 = -2\varepsilon$ as $q_i \cdot n_\varepsilon = 0$. Hence it has to vanish in the limit $\varepsilon \to 0$.

Summarizing we have shown that a general one-loop amplitude in a massless renormalizable four-dimensional quantum field theory can always be expressed in terms of scalar box, triangle and bubble integrals plus a rational part $\mathscr{R}$. Massive theories will also contain scalar tadpole integrals. Moreover, $\mathscr{R}$ only receives contributions from UV divergent tensor integrals whose numerators probe the regulating -2ε dimensions of the loop momentum.

3.4.4 An Example: The Photon Self-Energy in Massless QED

Let us work through an explicit simple example to see the discussed integrand reduction techniques at work. For this consider the photon self-energy graph in massless QED of Fig. 3.3.

Fig. 3.3 The one-loop correction to the photon propagator in massless QED

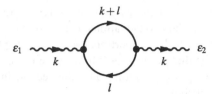

$$\Pi_{12} = -e^2 \int \frac{d^D l}{(2\pi)^D} \frac{\mathrm{Tr}(\not{\varepsilon}_1 (\not{k} + \not{l}) \not{\varepsilon}_2 \not{l})}{l^2 (l+k)^2}. \tag{3.71}$$

Here we have introduced polarization vectors ε_i at the external legs, but this is to be understood only as a place-holder to soak up the space-time indices. This one-loop correction to the propagator is to be inserted into larger diagrams and has off-shell photon legs with $k^2 \neq 0$. Ward identities demand the transversality of the photon self-energy, i.e. the vanishing of Π_{12} for $\varepsilon_i \to k$.

From the previous discussion it should be clear that Π_{12} may be written as a linear combination of bubbles Eq. (3.17) and tadpoles Eq. (3.16) along with a rational piece $\mathscr{R}$. The tadpoles vanish in dimensional regularization as the fermions in the loop are massless. The one-point functions also yield no contributions to the rational piece as is obvious from Eq. (3.54). Thus only the reduction of the two-point integral needs to be performed. This is a $d_p = 1$ and $d_t = 3 - 2\varepsilon$ dimensional integral for which we decompose the loop momentum l^μ as

$$l^\mu = (l \cdot k) v_1^\mu + l_\perp^\mu + (l \cdot n_\varepsilon) n_\varepsilon^\mu$$
$$\text{where } l_\perp^\mu = (l \cdot n_2) n_2^\mu + (l \cdot n_3) n_3^\mu + (l \cdot n_4) n_4^\mu \text{ and } k \cdot n_i = 0 = k \cdot n_\varepsilon. \tag{3.72}$$

One easily convinces oneself that $v_1^\mu = \frac{k^\mu}{k^2}$. In order to isolate the bubble coefficients of Π_{12} we put the two fermion propagators on-shell:

$$l^2 = 0, \qquad (l+k)^2 = 0. \tag{3.73}$$

These are solved for loop momenta of the form

$$l^\mu = -\frac{1}{2} k^\mu + l_\perp^\mu + (l \cdot n_\varepsilon) n_\varepsilon^\mu \quad \text{obeying} \quad l_\perp^2 + (l \cdot n_\varepsilon)^2 = -\frac{1}{4} k^2. \tag{3.74}$$

Hence the two conditions Eq. (3.73) fix the parallel components of l and determine the length of the transverse part. The idea is now to insert the on-shell l^μ found into the numerator of Eq. (3.71). This will directly yield the bubble coefficients as any off-shell components of l^μ would only give rise to terms proportional to l^2 or $(l+k)^2$ in the numerator which contribute to the vanishing tadpole integrals Eq. (3.16). Let us write $l^\mu = -\frac{1}{2} k^\mu + L^\mu$ and $k^\mu + l^\mu = +\frac{1}{2} k^\mu + L^\mu$ with $L^\mu = l_\perp^\mu + (l \cdot n_\varepsilon) n_\varepsilon^\mu$. The numerator then takes the on-shell form

$$N = \mathrm{Tr}\left[\not{\varepsilon}_1 \left(\frac{1}{2} \not{k} + \not{L} \right) \not{\varepsilon}_2 \left(-\frac{1}{2} \not{k} + \not{L} \right) \right]. \tag{3.75}$$

Since terms linear in L integrate to zero only two terms contribute to the integral

$$N = -\frac{1}{4} \text{Tr}[\not{\varepsilon}_1 \not{k} \not{\varepsilon}_2 \not{k}] + \text{Tr}[\not{\varepsilon}_1 \not{L} \not{\varepsilon}_2 \not{L}]. \tag{3.76}$$

Using the Dirac matrix trace identity $\text{Tr}[\not{\varepsilon}_1 \not{k} \not{\varepsilon}_2 \not{k}] = 4(2(\varepsilon_1 \cdot k)(\varepsilon_1 \cdot k) - (\varepsilon_1 \cdot \varepsilon_2)k^2)$ and $L^2 = -\frac{1}{4}k^2$ we deduce

$$N = 2k^2 \left(\varepsilon_1 \cdot \varepsilon_2 - \frac{(\varepsilon_1 \cdot k)(\varepsilon_2 \cdot k)}{k^2} \right) + 8(\varepsilon_1 \cdot l_\perp)(\varepsilon_2 \cdot l_\perp), \quad \text{as } \varepsilon_i \cdot n_\varepsilon = 0. \tag{3.77}$$

The first term in the above contributes only to the scalar bubble integral. The second term gives rise to the following integral

$$\int \frac{d^D l}{(2\pi)^D} \frac{(\varepsilon_1 \cdot l_\perp)(\varepsilon_2 \cdot l_\perp)}{(l_\parallel^2 + l_\perp^2 + l_\varepsilon^2)((k + l_\parallel)^2 + l_\perp^2 + l_\varepsilon^2)}, \quad \text{where } l_\parallel \parallel k. \tag{3.78}$$

Using the rotational symmetry in the transverse space as in Eq. (3.65) we can replace

$$(\varepsilon_1 \cdot l_\perp)(\varepsilon_2 \cdot l_\perp) \rightarrow \frac{1}{3}(\varepsilon_1^\perp \cdot \varepsilon_2^\perp)l_\perp^2$$

$$= \frac{1}{3}\left(\varepsilon_1 \cdot \varepsilon_2 - \frac{(\varepsilon_1 \cdot k)(\varepsilon_2 \cdot k)}{k^2} \right)\left(-\frac{1}{4}k^2 - (l \cdot n_\varepsilon)^2 \right) \tag{3.79}$$

in this integral. Putting everything together we find the total integral reduction of the photon self-energy Eq. (3.71)

$$\Pi_{12} = -\frac{4}{3}e^2 k^2 \left(\varepsilon_1 \cdot \varepsilon_2 - \frac{(\varepsilon_1 \cdot k)(\varepsilon_2 \cdot k)}{k^2} \right) \int \frac{d^D l}{(2\pi)^D} \frac{1}{l^2(l+k)^2}$$

$$+ \frac{8}{3}e^2 \left(\varepsilon_1 \cdot \varepsilon_2 - \frac{(\varepsilon_1 \cdot k)(\varepsilon_2 \cdot k)}{k^2} \right) \int \frac{d^D l}{(2\pi)^D} \frac{(l \cdot n_\varepsilon)^2}{l^2(l+k)^2}, \tag{3.80}$$

recovering transversality as we should. We note the following explicit results of the basis integrals for $\varepsilon \rightarrow 0$ (see [4, 9])

$$\int \frac{d^D l}{(2\pi)^D} \frac{1}{l^2(l+k)^2} = ic_\Gamma \frac{1}{(-k^2)^\varepsilon}\left(\frac{1}{\varepsilon} + 2 \right) + \mathscr{O}(\varepsilon),$$

$$\int \frac{d^D l}{(2\pi)^D} \frac{(l \cdot n_\varepsilon)^2}{l^2(l+k)^2} = ic_\Gamma \left(-\frac{k^2}{6} \right) + \mathscr{O}(\varepsilon), \tag{3.81}$$

where $c_\Gamma = \frac{1}{(4\pi)^{2-\varepsilon}} \frac{\Gamma(1+\varepsilon)\Gamma^2(1-\varepsilon)}{\Gamma(1-2\varepsilon)}$. Equation (3.80) is the complete result to all orders in ε. As was discussed at the end of the previous section the rational piece (the second term on Eq. (3.80)) arises as an effect of the regulating -2ε extra-dimension. The pole of Π_{12} comes exclusively from the first term and reads

$$\Pi_{12}|_{\text{pole}} = \frac{1}{\varepsilon}\frac{ie^2}{12\pi^2}k^2\left(\varepsilon_1 \cdot \varepsilon_2 - \frac{(\varepsilon_1 \cdot k)(\varepsilon_2 \cdot k)}{k^2} \right). \tag{3.82}$$

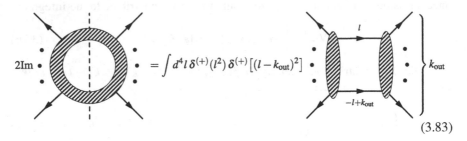

$$2\text{Im} \quad \bullet \quad = \int d^4 l \, \delta^{(+)}(l^2) \, \delta^{(+)}\big[(l - k_\text{out})^2\big] \quad \bullet \quad \Big\} k_\text{out}$$

(3.83)

Fig. 3.4 The optical theorem

3.5 Unitarity

We now turn to a modern and very efficient method to compute one-loop scattering amplitudes in gauge theories based on generalized unitarity. This principle represents a generalization of the optical theorem. Using the insights of the previous section it is clear that the task of computing a one-loop amplitude in gauge theories has been reduced to determining the coefficients $c_{n;j}$ multiplying the integrals in the general expansion of Eq. (3.15) along with finding the rational piece $\mathcal{R}$. The generalized unitarity method uses this knowledge by studying the discontinuities of an amplitude in various kinematical channels and exploiting the factorization property of the amplitude. The method allows for the determination of the coefficients $c_{n;j}$ in terms of on-shell phase space integrals over products of tree-level amplitudes. One is thus able to reconstruct one-loop amplitudes from tree-level data without the laborious Feynman diagram expansion.

To motivate this method let us recall the optical theorem in quantum field theory, which builds upon the fundamental property of the unitarity of the S-matrix,

$$SS^\dagger = \mathbb{1}. \tag{3.84}$$

We write $S = \mathbb{1} + iT$ as we did in chapter one, where T yields the non-trivial part of the scattering matrix. Using the perturbative expansion of T in the coupling constant g parametrizing the interaction strength we have

$$T = g^2 T^{(\text{tree})} + g^4 T^{(\text{1-loop})} + g^6 T^{(\text{2-loop})} + \cdots \tag{3.85}$$

starting out at order g^2 as a non-trivial part requires at least one interaction. The unitarity of $S = \mathbb{1} + iT$ implies the non-linear relation for T

$$-i\big(T - T^\dagger\big) = TT^\dagger, \tag{3.86}$$

known as the optical theorem. Importantly it relates contributions to T at different loop orders to one another. At the two lowest orders in the coupling constant g expansion Eq. (3.86) states

$$T^{(\text{tree})} = T^{(\text{tree})\dagger}, \qquad -i\big(T^{(\text{1-loop})} - T^{(\text{1-loop})\dagger}\big) = T^{(\text{tree})} T^{(\text{tree})}. \tag{3.87}$$

For the matrix elements of T between asymptotic states we write $T_{oi} = \langle \text{out}|T|\text{in}\rangle$. Consequently

$$\langle \text{out}|T^\dagger|\text{in}\rangle = \langle \text{in}|T|\text{out}\rangle^* = T_{io}^*. \tag{3.88}$$

Hence the second relation in Eq. (3.87) amounts to Cutkosky's rule [10]

$$-i\left(T_{oi}^{(1\text{-loop})} - T_{io}^{(1\text{-loop})*}\right) = \int d\mu T_{o\mu}^{(\text{tree})} T_{\mu i}^{(\text{tree})}, \tag{3.89}$$

where the integral $\int d\mu$ is symbolic for sums over particle species and helicities as well as phase-space integrals over all intermediate *on-shell* states $|\mu\rangle$ of the theory. Appealing to time-reversal invariance the l.h.s. of the above relation becomes twice the imaginary part of the one-loop amplitude, $2\,\text{Im}\,T_{oi}$. This is graphically represented in Fig. 3.4.

The analytic structure of amplitudes changes distinctively when one goes from tree- to loop-level. We saw that color ordered tree-amplitudes could only develop simple poles in the region momenta, i.e. sums of cyclically adjacent momenta. These are the isolated points of the amplitude when a virtual particle could go on-shell. In fact we used this feature extensively in the previous chapter. Loop-level amplitudes, on the other hand, are generically expressed in terms logarithms, dilogarithms and other special function of the kinematical invariants which can have branch cut singularities as complex functions of the kinematical invariants. The imaginary part of a one-loop amplitude arises from the discontinuity across such a branch cut singularity as a function of the kinematical invariants. For example the function $\log[s/\mu]$ has such a discontinuity $\text{Disc}(s)\log[s/\mu] = 2\pi i$ accross the s-channel branch cut.

This discontinuity may be traced back to the subspaces in the loop-momenta integration when a virtual particle goes on shell. Here the on-shell propagators become purely imaginary by the Feynman $i0$ prescription. One can think of this as "cutting" a propagator of the one-loop amplitude apart. Crossing the cut of a massless propagator with Feynman's $i0$ prescription gives rise to a principal and localized part via the distributional relation

$$\frac{1}{p^2 \pm i0} = P\left(\frac{1}{p^2}\right) \mp i\pi\delta(p^2), \tag{3.90}$$

applied to the virtual particle propagator. Here P denotes the principal part. Therefore the discontinuity of the one loop amplitude across a branch cut can be computed by replacing two propagators in a given channel by $\delta(p^2)$. In fact there is an important subtlety related to causality, which we will not discuss here. It implies the restriction to positive on-shell energies along the cut, i.e. one has the delta function $\delta^{(+)}(p^2) = \Theta(p^0)\delta(p^2)$ in Eq. (3.90).

3.5.1   Two-Particle Cuts of the Four Gluon Amplitude

Let us see this principle at work in a four-point gluon amplitude $A_4(1^-, 2^-, 3^+, 4^+)$ at the one-loop order. Generally we will have the expansion in terms of boxes, triangles and bubbles and a rational piece

$$A_4^{(1\text{-loop})}\left(1^-,2^-,3^+,4^+\right) = c_4 I_4 + \sum_{i_3} c_{3;i_3} I_{3;i_3} + \sum_{i_2} c_{2;i_2} I_{2;i_2} + \mathscr{R}. \tag{3.91}$$

If we now consider the discontinuity of this equation, then the l.h.s. may be rewritten using the optical theorem as a product of tree-level amplitudes integrated over the Lorentz invariant phase space and summed over all possible intermediate particle helicities. The r.h.s. on the other hand will only have discontinuities in the basis integrals I_4, I_3, I_2, as the coefficients $c_{n;j}$ are rational functions of the kinematical invariants and thus not affected by branch cut singularities. Similarly the rational piece R does not contribute to the discontinuity. One speaks of the non-rational contributions as being 'cut constructible'.

3.5.1.1 The s-Channel

To begin with let us consider a cut in the s-channel or $(1, 2)$-channel. Denoting the branch cut in this channel by $\mathrm{Disc}(s)$ we have

$$\sum_{h_1,h_2} \int d\mu A_4^{(\text{tree})}\left(1^-,2^-,l_1^{h_1},l_2^{h_2}\right) A_4^{(\text{tree})}\left(3^+,4^+,-l_2^{-h_2},-l_1^{-h_1}\right)$$

$$= c_4 \,\mathrm{Disc}(s) I_4 + \sum_{i_3} c_{3;i_3} \,\mathrm{Disc}(s) I_{3;i_3} + \sum_{i_2} c_{2;i_2} \,\mathrm{Disc}(s) I_{2;i_2}, \tag{3.92}$$

where the sums run over all helicities propagating through the channels and the Lorentz invariant phase space measure reads

$$d\mu = d^4 l_1 d^4 l_2 \delta^{(+)}\left(l_1^2\right) \delta^{(+)}\left(l_2^2\right) \delta^{(4)}\left(l_1 + l_2 + p_1 + p_2\right), \tag{3.93}$$

reflecting the double-cut by putting the intermediate legs l_1 and l_2 on-shell. The l.h.s. of Eq. (3.92) may be evaluated as follows. From Fig. 3.5 one sees that the helicities on the cut legs have to take the values $h_1 = h_2 = +1$ as the only non-vanishing tree-level amplitude involving two negative helicity gluons is the pure gluon amplitude $A_4(1^-, 2^-, l_1^{+1}, l_2^{+1})$. All other amplitudes obtained by replacing legs l_1 and l_2 by negative helicity gluons, scalars or fermions, i.e. putting helicities $h_{\{l_1,l_2\}} < +1$, vanish. Hence the integral of the l.h.s. of Eq. (3.92) reads

$$\mathrm{Disc}(s) A_4^{(1\text{-loop})} = \int d\mu A_4^{(\text{tree})}\left(1^-,2^-,l_1^+,l_2^+\right) A_4^{(\text{tree})}\left(3^+,4^+,-l_2^-,-l_1^-\right),$$

with $l_1 + l_2 = p_3 + p_4 = -p_1 - p_2$. \hfill (3.94)

Fig. 3.5 The s- and
t-channel cuts of the 4 gluon
amplitude
$A_4(1^-, 2^-, 3^+, 4^+)$

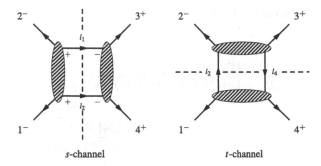

s-channel   t-channel

The integrand in the above takes the form

$$\frac{\langle 12 \rangle^4}{\langle 12 \rangle \langle 2l_1 \rangle \langle l_1 l_2 \rangle \langle l_2 1 \rangle} \frac{\langle l_1 l_2 \rangle^4}{\langle l_1 3 \rangle \langle 34 \rangle \langle 4l_2 \rangle \langle l_2 l_1 \rangle}$$

$$= -\frac{\langle 12 \rangle^4}{\langle 12 \rangle \langle 23 \rangle \langle 34 \rangle \langle 41 \rangle} \frac{\langle 23 \rangle \langle 41 \rangle \langle l_1 l_2 \rangle^2}{\langle 2l_1 \rangle \langle l_2 1 \rangle \langle l_1 3 \rangle \langle 4l_2 \rangle}, \tag{3.95}$$

where we have factored out the tree-level MHV$_4$ amplitude $A_4^{(\text{tree})}(1^-, 2^-, 3^+, 4^+)$.

One could proceed by performing the phase space integral in order to compare the result with the discontinuities of the known master integral functions on the r.h.s. of Eq. (3.92). However, there is an easier way to find the coefficients by directly comparing to the *integrands* arising in the r.h.s. of Eq. (3.92) through the s-channel cuts of the contributing box, triangle and bubble master-integrals. Note also that the integrands on the r.h.s. come with the same delta and theta-function contributions in $\delta^{(+)}(l_i^2)$ so that we can drop these in the comparison of integrands. Nevertheless, we have to keep the on-shell relations $l_i^2 = 0$ and $l_1 + l_2 = p_3 + p_4 = -p_1 - p_2$ in mind. The first contribution on the r.h.s. arises from the s-channel cut box integral

$$\text{Disc}(s)I_4 = $$

$$= \int d\mu \frac{1}{(l_2 + p_1)^2} \frac{1}{(l_1 - p_3)^2} = -\int d\mu \frac{1}{\langle l_2 1 \rangle [1l_2]} \frac{1}{\langle l_1 3 \rangle [3l_1]}. \tag{3.96}$$

Comparing this to the second term of the l.h.s. integrand in Eq. (3.95) shows that the quantities $\langle l_2 1 \rangle$ and $\langle l_1 3 \rangle$ already appear in the denominator. With some algebra the l_i dependence of the second term in Eq. (3.95) may be seen to equal the integrand of the cut box Eq. (3.96):

$$
-\frac{\langle 23\rangle\langle 41\rangle\langle l_1 l_2\rangle^2}{\langle 2l_1\rangle\langle l_2 1\rangle\langle l_1 3\rangle\langle 4l_2\rangle} = \frac{1}{(l_2+p_1)^2(l_1-p_3)^2}\frac{\langle 23\rangle\langle 41\rangle\overbrace{[3l_1]\langle l_1 l_2\rangle}^{[34]\langle 4l_2\rangle}\overbrace{[1l_2]\langle l_2 l_1\rangle}^{-[12]\langle 2l_1\rangle}}{\langle 2l_1\rangle\langle 4l_2\rangle}
$$

$$
= -\frac{\langle 23\rangle[34]\overbrace{\langle 41\rangle[12]}^{-\langle 43\rangle[32]}}{(l_2+p_1)^2(l_1-p_3)^2} = \frac{(p_2+p_3)^2(p_3+p_4)^2}{(l_2+p_1)^2(l_1-p_3)^2}
$$

$$
= \frac{st}{(l_2+p_1)^2(l_1-p_3)^2}, \tag{3.97}
$$

using the on-shell relations $l_1+l_2 = p_3+p_4 = -p_1-p_2$. Putting everything together we have shown that

$$
\mathrm{Disc}(s)A_4^{(\text{1-loop})}(1^-,2^-,3^+,4^+) = st A_4^{(\text{tree})}(1^-,2^-,3^+,4^+)\,\mathrm{Disc}(s)I_4, \tag{3.98}
$$

which in view of the master integral expansion Eq. (3.91) implies

$$
c_4 = st A_4^{(\text{tree})}(1^-,2^-,3^+,4^+). \tag{3.99}
$$

Moreover we also learn that there are no s-channel contributions from triangles and bubbles.

As we did not need to make any assumptions on the particle content and couplings of the theory to arrive at this result, we can state that in *any* gauge theory the one-loop gluon amplitude is

$$
A_4^{(\text{1-loop})}(1^-,2^-,3^+,4^+)
$$

$$
= st A_4^{(\text{tree})}(1^-,2^-,3^+,4^+)I_4(s,t) + \text{triangles} + \text{bubbles} + \mathscr{R}, \tag{3.100}
$$

where the only triangles and bubbles that may contribute are the three master integrals displayed in Fig. 3.6 above with non-vanishing t-channel cuts, with $s = s_{12}$ and $t = s_{23}$.

3.5.1.2 The t-Channel

Let us now turn to the remaining analysis of the t-channel cuts. The starting point is the equation

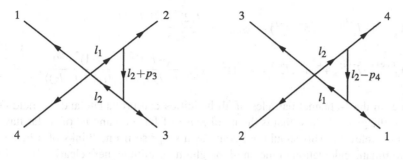

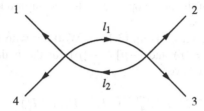

Fig. 3.6 The triangle and bubble master-integrals with t-channel cuts that can contribute to $A_4^{\text{1-loop}}(1^-, 2^-, 3^+, 4^+)$

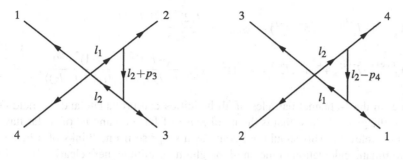

$$= \sum_{h_1, h_2} \int d\tilde{\mu} A_4^{(\text{tree})}\left(-l_1^{-h_1}, 2^-, 3^+, l_2^{h_2}\right) A_4^{(\text{tree})}\left(-l_2^{-h_2}, 4^+, 1^-, l_1^{h_1}\right)$$

$$= c_4 \, \text{Disc}(t) I_4 + \sum_{i_3 = \{a,b\}} c_{3;i_3} \, \text{Disc}(t) I_{3;i_3} + c_2 \, \text{Disc}(t) I_2, \tag{3.101}$$

where the sum over the helicity runs over $h_{1,2} = \{-1, -\frac{1}{2}, 0, \frac{1}{2}, 1\}$ and the Lorentz invariant phase space measure now is

$$d\tilde{\mu} = d^4 l_1 d^4 l_2 \delta^{(+)}\left(l_1^2\right) \delta^{(+)}\left(l_2^2\right) \delta^{(4)}(-l_1 + l_2 + p_2 + p_3). \tag{3.102}$$

We note the relevant 4-point tree amplitudes involving two gluons of opposite helicities from our discussion in Chap. 2,

$$A_4^{(\text{tree})}\left(-l_1^{-h_1}, 2^-, 3^+, l_2^{h_2}\right) = \delta_{h_1, h_2} \frac{\langle -l_1 2\rangle^{2+2h_1} \langle l_2 2\rangle^{2-2h_1}}{\langle -l_1 2\rangle \langle 23\rangle \langle 3 l_2\rangle \langle l_2(-l_1)\rangle},$$

$$A_4^{(\text{tree})}\left(-l_2^{-h_2}, 4^+, 1^-, l_1^{h_1}\right) = (-1)^{2h_1} \delta_{h_1, h_2} \frac{\langle -l_2 1\rangle^{2+2h_1} \langle l_1 1\rangle^{2-2h_1}}{\langle -l_2 4\rangle \langle 41\rangle \langle 1 l_1\rangle \langle l_1(-l_2)\rangle}.$$
(3.103)

Hence, in the t-channel particles of all helicities can propagate and the field content of the gauge theory, that is the number n_f of fermion and n_s of scalar flavors, becomes relevant. This should not come as a surprise if one thinks of a Feynman diagrammatic calculation of the one-loop gluon amplitude: here clearly all particles coupling to the gluon circulate in the off-shell loop. Rather the just discovered fact that the s-channel cut of the one-loop $(-, -, +, +)$ amplitude is insensitive to the field content is unexpected.

We now evaluate the r.h.s. of Eq. (3.101). Working with the negative momentum prescription $|-p\rangle = |p\rangle$ and $|-p] = -|p]$ for the helicity spinors ensuring $|-p]\langle -p| = -|p]\langle p|$ we have

$$\sum_h A_4^{(\text{tree})}\left(-l_1^{-h}, 2^-, 3^+, l_2^h\right) A_4^{(\text{tree})}\left(-l_2^{-h}, 4^+, 1^-, l_1^h\right)$$

$$= \sum_{h=\{-1, -\frac{1}{2}, 0, \frac{1}{2}, 1\}} (-1)^{2h} \frac{n_{|h|}(\langle l_1 2\rangle \langle l_2 1\rangle)^{2+2h} (\langle l_2 2\rangle \langle l_1 1\rangle)^{2-2h}}{\langle l_1 2\rangle \langle 23\rangle \langle 3 l_2\rangle \langle l_2 l_1\rangle \langle l_2 4\rangle \langle 41\rangle \langle 1 l_1\rangle \langle l_1 l_2\rangle},$$
(3.104)

with the multiplicities $n_1 = 1$, $n_{1/2} = n_f$ and $n_0 = n_s$. The sum in the numerator may be rewritten as

$$\underbrace{(\langle l_1 2\rangle \langle l_2 1\rangle - \langle l_2 2\rangle \langle l_1 1\rangle)^4}_{\langle 12\rangle^4 \langle l_1 l_2\rangle^4}$$

$$- (n_f - 4)\left(\langle l_1 2\rangle^2 \langle l_2 1\rangle^2 + \langle l_2 2\rangle^2 \langle l_1 1\rangle^2\right) \langle l_1 2\rangle \langle l_2 2\rangle \langle l_2 1\rangle \langle l_1 1\rangle$$

$$+ (n_s - 6) \langle l_1 2\rangle^2 \langle l_2 2\rangle^2 \langle l_2 1\rangle^2 \langle l_1 1\rangle^2.$$
(3.105)

3.5.1.3 Box Contributions

We see that something very special occurs for the case of $n_f = 4$ and $n_s = 6$, i.e. the field content of $\mathcal{N} = 4$ super Yang-Mills theory: then the entire sum in Eq. (3.105) reduces to a single term which after factorizing out the tree-level amplitude and by making use of a cyclic permutation of labels in Eq. (3.97) may be transformed to

$$A_4^{(\text{tree})}\left(1^-, 2^-, 3^+, 4^+\right) \frac{\langle 12\rangle \langle 34\rangle \langle l_1 l_2\rangle^2}{\langle 1 l_1\rangle \langle l_2 4\rangle \langle l_1 2\rangle \langle 3 l_2\rangle}$$

$$= A_4^{(\text{tree})}\left(1^-, 2^-, 3^+, 4^+\right) \frac{st}{(l_1 - p_2)^2 (l_2 - p_4)^2}.$$
(3.106)

But this is nothing else than the t-channel cut box integral multiplied by c_4! Hence in the case of $\mathcal{N} = 4$ super Yang-Mills we have shown that

$$\text{Disc}(t)A_{4,\mathcal{N}=4\,\text{SYM}}^{(1\text{-loop})}\left(1^-,2^-,3^+,4^+\right) = st A_4^{(\text{tree})}\left(1^-,2^-,3^+,4^+\right)\text{Disc}(t)I_4,$$
(3.107)

which in conjunction with Eq. (3.98) implies that there are no triangles or bubbles contributing in the full answer as they would have been detected in the t-channel. The potential triangles and bubbles hide in the terms canceled by the choice $n_f = 4$ and $n_s = 6$ in Eq. (3.105). We have thus proven the very compact one-loop result for $\mathcal{N} = 4$ super Yang-Mills

$$A_{4,\mathcal{N}=4\,\text{SYM}}^{(1\text{-loop})}\left(1^-,2^-,3^+,4^+\right) = st A_4^{(\text{tree})}\left(1^-,2^-,3^+,4^+\right)I_4(s,t),$$
(3.108)

modulo possible rational pieces. These may indeed be shown to vanish as well.

3.5.1.4 Triangle Contributions

Let us now see what can be said about the remaining terms for generic n_f and n_s in Eq. (3.105). These should equal the t-channel cuts of the one-mass triangles (a) and (b) as well as the bubble graph of Fig. 3.6

$$c_{3;a}\,\text{Disc}(t)I_{3;a} + c_{3;b}\,\text{Disc}(t)I_{3;b} + c_2\,\text{Disc}(t)I_2$$

$$= \frac{1}{\langle 23\rangle\langle 41\rangle}\frac{\langle l_2 1\rangle\langle l_2 2\rangle}{\langle 3l_2\rangle\langle l_2 4\rangle}\frac{1}{\langle l_1 l_2\rangle^2}\Big[(4 - n_f)\big(\langle l_1 1\rangle^2\langle l_2 2\rangle^2 + \langle l_1 2\rangle^2\langle l_2 1\rangle^2\big)$$

$$+ (n_s - 6)\langle l_1 1\rangle\langle l_2 2\rangle\langle l_1 1\rangle\langle l_2 2\rangle\Big].$$
(3.109)

In order to isolate the poles of the t-channel cut triangles (a) and (b) we first use Schouten's identity on the second factor of the r.h.s.: $\langle l_2 1\rangle\langle 34\rangle - \langle l_2 4\rangle\langle 31\rangle + \langle l_2 3\rangle\langle 14\rangle$ to find

$$\frac{\langle l_2 1\rangle\langle l_2 2\rangle}{\langle 3l_2\rangle\langle l_2 4\rangle} = \frac{1}{\langle 34\rangle}\left(\frac{\langle 31\rangle\langle l_2 2\rangle}{\langle 3l_2\rangle} + \frac{\langle 41\rangle\langle l_2 2\rangle}{\langle l_2 4\rangle}\right)$$

$$= \frac{1}{\langle 34\rangle}\left(\frac{\langle 13\rangle\langle 2|l_2|3]}{(l_2 + p_3)^2} + \frac{\langle 14\rangle\langle 2|l_2|4]}{(l_2 - p_4)^2}\right),$$
(3.110)

where we lifted to propagators in the last step using the relations

$$\frac{1}{\langle 3l_2\rangle} = \frac{[l_2 3]}{(l_2 + p_3)^2}, \qquad \frac{1}{\langle l_2 4\rangle} = \frac{[l_2 4]}{(l_2 - p_4)^2}.$$
(3.111)

Before inserting Eq. (3.110) into Eq. (3.109) we furthermore rewrite the pole in $\langle l_1 l_2\rangle^2$ in Eq. (3.109) with the help of

$$\frac{1}{\langle l_1 l_2\rangle^2} = \frac{[l_1 l_2]^2}{t^2}.$$
(3.112)

Pulling the emerging $[l_1 l_2]^2$ term into the square bracket expression in Eq. (3.109) allows for a complete rewriting of these terms through the four-momentum l_2^μ. For this we use recalling $l_1 - l_2 = p_2 + p_3 = -p_1 - p_4$

$$[l_1 l_2]\langle l_1 1 \rangle = [l_2 4]\langle 41 \rangle, \qquad [l_1 l_2]\langle l_1 2 \rangle = -[l_2 3]\langle 32 \rangle, \qquad (3.113)$$

to finally arrive at

$$c_{3;a}\operatorname{Disc}(t)I_{3;a} + c_{3;b}\operatorname{Disc}(t)I_{3;b} + c_2\operatorname{Disc}(t)I_2$$

$$= \frac{1}{\langle 23 \rangle \langle 34 \rangle \langle 41 \rangle}\frac{1}{t^2}\left(\frac{\langle 13 \rangle \langle 2|l_2|3]}{(l_2 + p_3)^2} + \frac{\langle 14 \rangle \langle 2|l_2|4]}{(l_2 - p_4)^2}\right)$$

$$\times \left[(4 - n_f)\big(\langle 41 \rangle^2 \langle 2|l_2|4]^2 + \langle 23 \rangle^2 \langle 1|l_2|3]^2\big)\right.$$

$$\left. + (n_s - 6)\langle 23 \rangle \langle 41 \rangle \langle 2|l_2|4]\langle 1|l_2|3]\right]. \qquad (3.114)$$

Through these manipulations we have rewritten the r.h.s. in a fashion allowing for a lift off the cuts by taking the l_2 loop-momentum off-shell and including the cut propagators for the two triangle cuts seperately. For this we may lift the cuts according to the prescriptions

$$(a): \quad \frac{\mathcal{N}_a(l_2)}{(l_2 + p_3)^2} \rightarrow \int \frac{d^4 l}{(2\pi)^4}\frac{\mathcal{N}_a(l)}{l^2(l + p_2 + p_3)^2(l + p_3)^2},$$

$$(b): \quad \frac{\mathcal{N}_b(l_2)}{(l_2 - p_4)^2} \rightarrow \int \frac{d^4 l}{(2\pi)^4}\frac{\mathcal{N}_b(l)}{l^2(l - p_1 - p_4)^2(l - p_4)^2}, \qquad (3.115)$$

where the momenta l on the r.h.s. are taken to be the off-shell loop momenta. One now has two distinct tensor three-point integrals and is in a position to perform an integral reduction along the lines discussed in Sect. 3.4 to reduce to the scalar triangle integral. It should be noted, however, that in performing this lift we are necessarily blind to any terms in $\mathcal{N}_a(l)$ proportional to the cut propagators l^2 and $(l + p_2 + p_3)^2$ respectively. Such terms would bring us down to bubble graphs, hence the lift Eq. (3.115) does capture all triangle contributions.

Starting with the lift of the (a) type triangle we are facing the tensor three-point integral

$$\frac{\langle 13 \rangle}{t^2 \langle 23 \rangle \langle 34 \rangle \langle 41 \rangle}\int \frac{d^4 l}{(2\pi)^4}\frac{\mathcal{N}_a(l)}{l^2(l + q_1)^2(l + q_2)^2}, \qquad (3.116)$$

$$q_1 = p_3, \qquad q_2 = p_2 + p_3,$$

with the numerator polynomial

$$\mathcal{N}_a(l) = \langle 2|l|3]\big[(4 - n_f)\big(\langle 41 \rangle^2 \langle 2|l_2|4]^2 + \langle 23 \rangle^2 \langle 1|l_2|3]^2\big)$$

$$+ (n_s - 6)\langle 23 \rangle \langle 41 \rangle \langle 2|l_2|4]\langle 1|l_2|3]\big]. \qquad (3.117)$$

Following the discussion in Sect. 3.4 we know that this numerator may be brought
into the form (compare Eq. (3.52))

$$\mathscr{N}_a(l) = \tilde{c}_0^a + \tilde{c}_1^a(l \cdot n_3) + \tilde{c}_2^a(l \cdot n_4) + \tilde{c}_4^a\left[(l \cdot n_3)^2 - (l \cdot n_4)^2\right]$$

$$+ \tilde{c}_5^a(l \cdot n_3)(l \cdot n_4) + \tilde{c}_6^a(l \cdot n_3)^3 + \tilde{c}_7^a(l \cdot n_4)^3 + \mathcal{O}(d_i), \qquad (3.118)$$

where we have dropped extra-dimensional components $(l \cdot n_\varepsilon)$ as we require the loop
momentum to be strictly four dimensional in the present discussion. This is justified
as we are not chasing the rational piece R presently. Note that generically terms
proportional to the inverse propagators $d_0 = l^2$, $d_1 = (l + q_1)^2$ and $d_2 = (l + q_2)^2$
are unknown in the above and the loop momentum is to be expanded as

$$l^\mu = (l \cdot q_1)v_1^\mu + (l \cdot q_2)v_2^\mu + (l \cdot n_3)n_3^\mu + (l \cdot n_4)n_4^\mu. \qquad (3.119)$$

Our goal is to find the coefficient $\tilde{c}_0^a$. A good strategy is to put *all* three propagators
on-shell, i.e. $d_i = 0$, as this removes the redundancy in the $\mathcal{O}(d_i)$ terms. There will
be two one-parameter families of solutions for these three on-shell conditions which
we shall call $\bar{l}^\mu$:

$$d_i(\bar{l}) = 0, \quad i = 0, 1, 2$$

$$\Leftrightarrow \quad \bar{l}^2 = 0, \qquad 2(\bar{l} \cdot q_1) = 0, \qquad 2(\bar{l} \cdot q_2) + q_2^2 = 0. \qquad (3.120)$$

Indeed with $q_2^2 = t$ we see that there are two solution $\bar{l}_\pm^\mu$ of the form

$$\bar{l}_\pm^\mu(\alpha) = -\frac{t}{2}v_2^\mu \pm \sqrt{-\alpha^2 - \frac{t^2}{4}v_2^2 n_3^\mu + \alpha n_4^\mu}. \qquad (3.121)$$

Noting that the dual basis vectors to q_1 and q_2 are $v_1 = \frac{2}{t}(p_2^\mu - p_3^\mu)$ and $v_2^\mu = \frac{2}{t}p_3^\mu$
respectively, we arrive at the final on-shell solutions

$$\bar{l}_\pm^\mu(\alpha) = -p_3^\mu \pm i\alpha n_3^\mu + \alpha n_4^\mu. \qquad (3.122)$$

In order to detect the scalar coefficient $\tilde{c}_0^a$ from equating Eqs. (3.117) and (3.118)
for the on-shell $\bar{l}_\pm(\alpha)$ we chose $\alpha = 0$ in the solution in order to switch off all
components parallel to $n_{3/4}^\mu$. Note that the coefficient $\tilde{c}_0^a$ cannot depend on the choice
of α. Nicely the result then trivializes

$$c_0^a = \mathscr{N}_a(l = -p_3) = \langle 2|p_3|3][\cdots] = 0. \qquad (3.123)$$

Thus, there are no type (a) triangle contributions! A similar argument shows that
also the (b) triangle contribution vanishes.

We have thus shown that for a generic gauge theory the $(--++)$ amplitude has
no triangle contributions

$$A_4^{(1\text{-loop})}\left(1^-, 2^-, 3^+, 4^+\right)$$

$$= st A_4^{(\text{tree})}\left(1^-, 2^-, 3^+, 4^+\right)I_4(s, t) + \text{bubbles} + \mathscr{R}. \qquad (3.124)$$

We note in closing that this is special to the 4-gluon case with two adjacent negative helicity gluons. For the five gluon amplitude $(--+++)$ at one-loop there are triangle contributions which may be found with the techniques exposed. This is due to the fact that only two-mass triangle integrals contribute which can arise only for $n > 4$ gluons.

3.5.2 Generalized Unitarity and Higher-Order Cuts

In the previous section we have seen how cutting two propagators in a one-loop amplitude in various channels allows one to construct the amplitude completely up to rational terms. The idea behind the generalized unitarity method to be discussed now is to go beyond double cuts and to construct the one-loop amplitude by applying also triple and quadruple cuts. Here one starts from the maximal number of cuts of the amplitude in all possible channels to find the corresponding box coefficients in terms of four tree-level amplitudes. From this one proceeds to triple, double and possibly single cuts to reconstruct the lower integral coefficients always subtracting the 'pollution' from higher order cuts at every step.

A number of different strategies have been devised to find the master integral coefficients using this method. We shall discuss here the efficient strategy of generalized unitarity applied to the integrand reduction discussed in Sect. 3.4. As we saw there the one-loop integrand with loop momentum l for a given one-loop amplitude reads

$$A_n^{\text{1-loop integrand}}(l) = \sum_{1 \leq i_1 < i_2 < i_3 < i_4 \leq n} \frac{d_{i_1 i_2 i_3 i_4}(l)}{d_{i_1} d_{i_2} d_{i_3} d_{i_4}} + \sum_{1 \leq i_1 < i_2 < i_3 \leq n} \frac{c_{i_1 i_2 i_3}(l)}{d_{i_1} d_{i_2} d_{i_3}}$$

$$+ \sum_{1 \leq i_1 < i_2 \leq n} \frac{b_{i_1 i_2}(l)}{d_{i_1} d_{i_2}} + \sum_{1 \leq i_1 \leq n} \frac{a_{i_1}(l)}{d_{i_1}}. \tag{3.125}$$

Confining oneself to four dimensions and working in the van Neerven-Vermaseren basis the numerators were shown to take the form (in a condensed notation)

$$d_{i_1 i_2 i_3 i_4}(l) = \tilde{d}_0 + \tilde{d}_1(l \cdot n_4), \tag{3.126}$$

$$c_{i_1 i_2 i_3}(l) = \tilde{c}_0 + \tilde{c}_1(l \cdot n_3) + \tilde{c}_2(l \cdot n_4) + \tilde{c}_4\big[(l \cdot n_3)^2 - (l \cdot n_4)^2\big]$$

$$+ \tilde{c}_5(l \cdot n_3)(l \cdot n_4) + \tilde{c}_6(l \cdot n_3)^3 + \tilde{c}_7(l \cdot n_4)^3, \tag{3.127}$$

$$b_{i_1 i_2}(l) = \tilde{b}_0 + \tilde{b}_1(l \cdot n_2) + \tilde{b}_2(l \cdot n_3) + \tilde{b}_3(l \cdot n_4) + \tilde{b}_5\big[(l \cdot n_2)^2 - (l \cdot n_4)^2\big]$$

$$+ \tilde{b}_6\big[(l \cdot n_3)^2 - (l \cdot n_4)^2\big] + \tilde{b}_7(l \cdot n_2)(l \cdot n_3)$$

$$+ \tilde{b}_8(l \cdot n_2)(l \cdot n_4) + \tilde{b}_9(l \cdot n_3)(l \cdot n_4), \tag{3.128}$$

$$a_{i_1}(l) = \tilde{a}_0 + \tilde{a}_1(l \cdot n_1) + \tilde{a}_2(l \cdot n_2) + \tilde{a}_3(l \cdot n_3) + \tilde{a}_4(l \cdot n_4). \tag{3.129}$$

The coefficients we are interested in are the $\tilde{d}_0$, $\tilde{c}_0$, $\tilde{b}_0$ and $\tilde{a}_0$ as the remaining terms in the above vanish upon integration over l. In the language of the integrand of Eq. (3.125), a unitarity cut corresponds to setting one inverse propagator to zero, $d_i = 0$ and removing the pole from the expression. Using suitable unitarity cuts we may isolate individual terms in Eq. (3.125).

One begins with all the maximal quadruple cuts isolating a single box term $d_{i_1 i_2 i_3 i_4}$. Here we have (we consider massless particles only for simplicity) the cut propagators in the nomenclature of Sect. 3.4

$$d_0 = l^2 = 0, \qquad d_1 = (l + q_1)^2 = 0,$$
$$d_2 = (l + q_2)^2 = 0, \qquad d_3 = (l + q_3)^2 = 0. \tag{3.130}$$

The last three equations tell us that $2l \cdot q_i = -q_i^2$. Hence the van Neerven-Vermaseren basis decomposition of l^μ for the physical dimension $d_p = 3$ of the box-integral reads

$$l^\mu = -\frac{1}{2} \sum_{i=1}^{3} q_i^2 v_i^\mu + (l \cdot n_4) n_4^\mu. \tag{3.131}$$

The coefficient $(l \cdot n_4)$ follows from the first equation in (3.130)

$$l^2 = 0 = \frac{1}{4} \left(\sum_{i=1}^{3} q_i^2 v_i^\mu \right)^2 + (l \cdot n_4)^2$$

$$\Rightarrow \quad (l \cdot n_4) = \pm \frac{1}{2} \sqrt{-\left(q_1^2 v_1^\mu + q_2^2 v_2^\mu + q_3^2 v_3^\mu \right)^2}. \tag{3.132}$$

We thus see that the quadruple cut freezes the loop momentum on two four-momenta

$$\bar{l}_\pm^\mu = -\frac{1}{2} \sum_{i=1}^{3} q_i^2 v_i^\mu \pm \frac{1}{2} \sqrt{-\left(q_1^2 v_1^\mu + q_2^2 v_2^\mu + q_3^2 v_3^\mu \right)^2} n_4^\mu, \tag{3.133}$$

which are generically complex. We now evaluate the left and right hand sides of the integrand relation Eq. (3.125) on the quadruple cuts $l = \bar{l}_\pm$. For the l.h.s. we find

$$A_n^{\text{1-loop integrand}}(l)\Big|_{\text{quadruple cut}, \, l = \bar{l}_\pm}$$
$$= A_1^{\text{tree}}(\bar{l}_\pm) A_2^{\text{tree}}(\bar{l}_\pm) A_3^{\text{tree}}(\bar{l}_\pm) A_4^{\text{tree}}(\bar{l}_\pm). \tag{3.134}$$

The r.h.s. of Eq. (3.125) on the cut simply is equal to $d_{i_1 i_2 i_3 i_4}(\bar{l}_\pm)$. With the expansion Eq. (3.126) this tells us that upon defining $D_\pm := A_1^{\text{tree}}(\bar{l}_\pm) A_2^{\text{tree}}(\bar{l}_\pm) A_3^{\text{tree}}(\bar{l}_\pm) \times$

$A_4^{\text{tree}}(\bar{l}_\pm)$ we have

$$D_\pm = \tilde{d}_0 \pm \tilde{d}_1 \frac{1}{2}\sqrt{-(q_1^2 v_1^\mu + q_2^2 v_2^\mu + q_3^2 v_3^\mu)^2}, \tag{3.135}$$

resulting in the solutions for the sought after box coefficients

$$\tilde{d}_0 = \frac{1}{2}(D_+ + D_-), \tag{3.136}$$

$$\tilde{d}_1 = \frac{1}{2}\frac{D_+ - D_-}{\sqrt{-(q_1^2 v_1^\mu + q_2^2 v_2^\mu + q_3^2 v_3^\mu)^2}}. \tag{3.137}$$

Hence the box coefficients are straightforwardly constructed from the fusion of four tree-level amplitudes. d_0 is the desired coefficient of the scalar box integral, the second coefficient d_1 will be needed for the determination of the triangle coefficients, although it does not contribute to the integral of Eq. (3.125).

For the triangle coefficients a triple cut is employed in order to isolate the particular triangle contribution one is interested in. However, one now has to take into account that the triple cuts of the box integral will also contribute and pollute the result. To overcome this problem one substracts off the box-contributions which are known from the quadruple cuts. Hence in order to find the seven triangle coefficients $\tilde{c}_i$ of Eq. (3.127) we do not consider $A_1^{\text{tree}}(\bar{l})A_2^{\text{tree}}(\bar{l})A_3^{\text{tree}}(\bar{l})$ on the triple cut momentum solutions $\bar{l}$ but rather the subtracted expression

$$A_{i_1}^{\text{tree}}(\bar{l})A_{i_2}^{\text{tree}}(\bar{l})A_{i_3}^{\text{tree}}(\bar{l}) - \sum_{i_4}\frac{d_{i_1 i_2 i_3 i_4}(\bar{l})}{d_{i_4}}. \tag{3.138}$$

We have seen a very similar procedure in the discussion of the previous subsection. Clearly the triple cut condition on the loop-momentum being a four-vector will lead to a one-parameter family of solutions. In fact we find two independent one-parameter families of triple cut momenta given by

$$\bar{l}_\pm^\mu(\alpha) = -\frac{1}{2}(q_1^2 v_1^\mu + q_2^2 v_2^\mu) \pm \frac{1}{2}\sqrt{-(q_1^2 v_1^\mu + q_2^2 v_2^\mu)^2}(\cos(\alpha)n_3^\mu + \sin(\alpha)n_4^\mu), \tag{3.139}$$

with $\alpha \in \mathbb{R}$. They obey the triple-cut conditions

$$d_0 = l^2 = 0, \qquad d_1 = (l + q_1)^2 = 0, \qquad d_2 = (l + q_2)^2 = 0. \tag{3.140}$$

In order to find the seven triangle coefficients one then considers the equation

$$A_{i_1}^{\text{tree}}(\bar{l}) A_{i_2}^{\text{tree}}(\bar{l}) A_{i_3}^{\text{tree}}(\bar{l}) - \sum_{i_4} \frac{d_{i_1 i_2 i_3 i_4}(\bar{l})}{d_{i_4}}$$

$$= \tilde{c}_0 + \tilde{c}_1 (\bar{l} \cdot n_3) + \tilde{c}_2 (\bar{l} \cdot n_4) + \tilde{c}_4 \big[(\bar{l} \cdot n_3)^2 - (\bar{l} \cdot n_4)^2 \big]$$

$$+ \tilde{c}_5 (\bar{l} \cdot n_3)(\bar{l} \cdot n_4) + \tilde{c}_6 (\bar{l} \cdot n_3)^3 + \tilde{c}_7 (\bar{l} \cdot n_4)^3 \qquad (3.141)$$

for arbitrary seven choices of $\bar{l}^\mu = \bar{l}_\pm^\mu(\alpha)$ e.g. by varying the parameter α. The resulting set of linear equations for the c_i may then be straightforwardly solved. In fact this procedure is very suitable also for a numerical implementation. Again c_0 is the desired scalar triangle-integral coefficients, while the other six coefficients are needed for the determination of the bubble coefficient.

By now the strategy should be clear. In order to find the bubble coefficients one considers the double cuts of the triangle and box subtracted expressions

$$A_{i_1}^{\text{tree}}(\bar{l}) A_{i_2}^{\text{tree}}(\bar{l}) - \sum_{i_3} \frac{c_{i_1 i_2 i_3}(\bar{l})}{d_{i_3}} - \frac{1}{2!} \sum_{i_3, i_4} \frac{d_{i_1 i_2 i_3 i_4}(\bar{l})}{d_{i_3} d_{i_4}}. \qquad (3.142)$$

This expression then equals the nine $\tilde{b}_i$ coefficients on the right-hand-side of Eq. (3.128). The double cut loop momenta obeying

$$d_0 = l^2 = 0, \qquad d_1 = (l + q_1)^2 = 0, \qquad (3.143)$$

are given by two-parameter families of solutions $\bar{l}_\pm(\alpha_1, \alpha_2)$

$$\bar{l}_\pm(\alpha_1, \alpha_2) = -\frac{1}{2} q_1^2 v_1^\mu \pm \sqrt{-\alpha_1^2 - \alpha_2^2 - \frac{1}{4}(q_1^2)^2 v_1^2 n_2^2} \, n_2^\mu$$

$$+ \alpha_1 n_3^\mu + \alpha_2 n_4^\mu, \qquad \alpha_i \in \mathbb{R}. \qquad (3.144)$$

Again one may solve for the nine remaining bubble coefficients $\tilde{b}_i$ by plugging arbitrary values of $\bar{l}_\pm(\alpha_1, \alpha_2)$ into the set of linear equations relating Eq. (3.142) to the r.h.s. of Eq. (3.128).

This completes our general discussion of the generalized unitarity principle to determine the integral coefficients of a one-loop amplitude. The construction has been brought into an algorithmic form in which no integrals need to be performed and only the knowledge of the tree-amplitudes is needed. They are in a sense recycled to generate the one-loop amplitudes. This procedure is also suited for a purely numerical implementation, which has been done in the literature. We now turn to the discussion of an explicit analytical example.

3.5.2.1 Example: Box Coefficients of the Split Helicity 5-Gluon Amplitude in Pure Yang-Mills

Let us study generalized unitarity through an explicit example and construct the box-coefficients for the five-gluon one-loop amplitude with helicity configuration

Fig. 3.7 The quadruple cut of the contribution $d_{12}I(s_{12})$ to the 5-gluon amplitude. The SW vertex is of MHV$_4$ type which fixes the helicities of its inner outgoing-legs to be $+1$

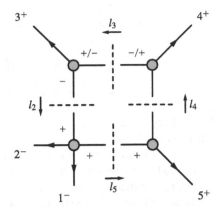

$A(1^-, 2^-, 3^+, 4^+, 5^+)$ in *pure* Yang-Mills theory. As the number of external legs is $n = 5$ the quadruple cut box will factor into one four-point amplitude and three three-point amplitudes. We may then write the decomposition

$$A_{5,\,\text{boxes}}^{\text{1-loop}} = d_{12}I(s_{12}) + d_{23}I(s_{23}) + d_{34}I(s_{34}) + d_{45}I(s_{45}) + d_{51}I(s_{51}), \quad (3.145)$$

where the introduced box-integral $I(s_{ij})$ has the massive leg $2p_i \cdot p_j$ arising from the two inflowing null momenta p_i and p_j attached to the four-point corner. For example $I(s_{12})$ reads explicitly

$$I(s_{12}) = \int \frac{d^4l}{(2\pi)^4} \frac{1}{d_0 d_2 d_3 d_4},$$

$$\text{with } d_0 = l^2, d_i = (l + q_i)^2, q_i = \sum_{j=1}^{i} p_j, \quad (3.146)$$

compare Fig. 3.7. The antisymmetry of the factorized amplitude under reflection $(12345) \to (21543)$ may be used to relate the coefficients d_{51} to d_{23} and d_{45} to d_{34} in Eq. (3.145) as we shall see shortly. Hence we only need to determine the three coefficients d_{12}, d_{23} and d_{34}.

We begin with d_{12}. Looking at the quadruple cut, the SW four-point amplitude in Fig. 3.7 needs to be of MHV$_4$ type and forces the outgoing helicities along the l_5 and $-l_2$ momentum legs to be $+1$. The remaining NW, NE and SE amplitudes are necessarily three-point types which forces the on-shell loop momenta to be complex. Looking at the NW and NE amplitudes we have the choices:

$$\begin{array}{llll} \text{NW:} & \text{MHV}_3\text{:}\ \tilde{\lambda}_{l_2} \sim \tilde{\lambda}_3 \sim \tilde{\lambda}_{l_3}, & \overline{\text{MHV}}_3\text{:}\ \lambda_{l_2} \sim \lambda_3 \sim \lambda_{l_3}, \\ \text{NE:} & \text{MHV}_3\text{:}\ \tilde{\lambda}_{l_3} \sim \tilde{\lambda}_4 \sim \tilde{\lambda}_{l_4}, & \overline{\text{MHV}}_3\text{:}\ \lambda_{l_3} \sim \lambda_4 \sim \lambda_{l_4}. \end{array} \quad (3.147)$$

As a matter of fact only two of these four choices are allowed: if we took a combination of identical helicity type for the NW and NE three-point amplitudes, i.e. a

Fig. 3.8 The two choices of Eq. (3.148) where *white* (*black*) blobs denote $\overline{\text{MHV}}_3$ (MHV_3) vertices. Note that the second choice MHV_3–$\overline{\text{MHV}}_3$–MHV_3 is inconsistent

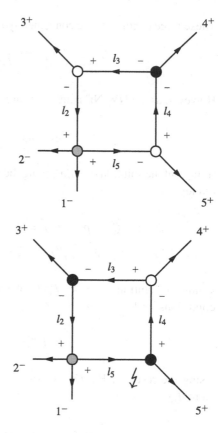

MHV$_3$–MHV$_3$ or $\overline{\text{MHV}}_3$–$\overline{\text{MHV}}_3$ combination, we would have $\lambda_3 \sim \lambda_4$ or $\tilde{\lambda}_3 \sim \tilde{\lambda}_4$ respectively, entailing $s_{34} = 2p_3 \cdot p_4 - 0$ which is not true for general momentum configurations. Hence we may only have a string of MHV$_3$–$\overline{\text{MHV}}_3$ or $\overline{\text{MHV}}_3$–MHV$_3$ combinations for the NW–NE amplitudes. This argument generalizes, hence only an alternating series of successive MHV$_3$ and $\overline{\text{MHV}}_3$ amplitudes are consistent. For our problem this leaves us with the two choices for the NW–NE–SE combinations

NW–NE–SE: $\quad \overline{\text{MHV}}_3 - \text{MHV}_3 - \overline{\text{MHV}}_3 \quad$ or $\quad \text{MHV}_3 - \overline{\text{MHV}}_3 - \text{MHV}_3$.

$$(3.148)$$

Upon studying the possible helicity configurations in Fig. 3.8 one convinces oneself that in the end only the first choice in Eq. (3.148) corresponding to the first graph of Fig. 3.8 is consistent.

Let us now find the on-shell loop momenta for this configuration. We have, using the notation of Fig. 3.7,

$$l_3^2 = 0, \qquad (l_3 - p_3)^2 = 2l_3 \cdot p_3 = 0, \qquad (l_3 + p_4)^2 = 2l_3 \cdot p_4 = 0. \quad (3.149)$$

These three conditions are conveniently solved by the two l_3's

$$\bar{l}_{3+}^{\alpha\dot{\alpha}} = \xi_+\lambda_3^\alpha\tilde{\lambda}_4^{\dot{\alpha}}, \qquad \bar{l}_{3-}^{\alpha\dot{\alpha}} = \xi_-\lambda_4^\alpha\tilde{\lambda}_3^{\dot{\alpha}}, \quad \xi_\pm \in \mathbb{C}. \tag{3.150}$$

However, as the NW-NE amplitudes are of $\overline{\text{MHV}}_3$–MHV_3 type we have

$$\lambda_{l_3} \sim \lambda_3 \quad \text{and} \quad \tilde{\lambda}_{l_3} \sim \tilde{\lambda}_4, \tag{3.151}$$

ruling out the solution $\bar{l}_{3-}$. Enforcing the remaining on-shell condition for the southern leg

$$(l_3 + p_4 + p_5)^2 = 0 = 2l_3 \cdot (p_4 + p_5) + 2p_4 \cdot p_5$$
$$\Rightarrow \quad \bar{l}_{3+}{}^\alpha{}_{\dot\alpha}\lambda_{5\alpha}\tilde{\lambda}_5^{\dot\alpha} = \langle 45\rangle[54], \tag{3.152}$$

yields the solution $\xi_+ = \langle 45\rangle/\langle 53\rangle$. We find only one solution to the quadruple cut conditions for l_3

$$l_3^{\alpha\dot{\alpha}} = -\frac{\langle 45\rangle}{\langle 35\rangle}\lambda_3^\alpha\tilde{\lambda}_4^{\dot{\alpha}}. \tag{3.153}$$

Using the result of Eq. (3.136) (with vanishing D_-) we find for the box coefficient d_{12}

$$d_{12} = \frac{1}{2}A_4^{\text{tree}}\left(1^-,2^-,-l_2^+,l_5^+\right)A_3^{\text{tree}}\left(3^+,-l_3^+,l_2^-\right)$$
$$\times A_3^{\text{tree}}\left(-l_4^-,l_3^-,4^+\right)A_3^{\text{tree}}\left(l_4^+,5^+,-l_5^-\right)$$
$$= \frac{1}{2}\frac{\langle 12\rangle^3}{\langle 2l_2\rangle\langle l_2l_5\rangle\langle l_51\rangle}\frac{[3l_3]^3}{[l_3l_2][l_23]}\frac{\langle l_4l_3\rangle^3}{\langle l_34\rangle\langle 4l_4\rangle}\frac{[l_45]^3}{[5l_5][l_5l_4]}$$
$$= -\frac{1}{2}\frac{\langle 12\rangle^3[3|l_3l_4|5]^3}{\langle 2|l_2|3]\langle 1|l_5l_4|4\rangle\langle 4|l_3l_2l_5|5]}. \tag{3.154}$$

Using the solution Eq. (3.153) and the relations $l_4 = l_3 + p_4$, $l_2 = l_3 - p_3$, $l_5 = l_3 + p_4 + p_5$ one finds via straightforward algebra

$$[3|l_3l_4|5] = -\frac{s_{45}s_{34}}{\langle 35\rangle}, \qquad \langle 2|l_2|3] = -\frac{\langle 45\rangle}{\langle 35\rangle}\langle 23\rangle[43],$$

$$\langle 1|l_5l_4|4\rangle = s_{45}\frac{\langle 51\rangle\langle 34\rangle}{\langle 35\rangle}, \qquad \langle 4|l_3l_2l_5|5] = s_{34}s_{45}\frac{\langle 34\rangle}{\langle 35\rangle}. \tag{3.155}$$

Inserting these into Eq. (3.154) yields the nice and compact result

$$d_{12} = -\frac{1}{2}s_{34}s_{45}\frac{\langle 12\rangle^3}{\langle 23\rangle\langle 34\rangle\langle 45\rangle\langle 51\rangle} = \frac{i}{2}s_{34}s_{45}A_5^{\text{tree}}(1^-,2^-,3^+,4^+,5^+). \quad (3.156)$$

The evaluation of the remaining four coefficients proceeds analogously and is performed in the following exercise. In all four cases there is again only one configuration of the four factorized amplitudes allowed. One finds in analogy to Eq. (3.156) in an on-shell diagram notation

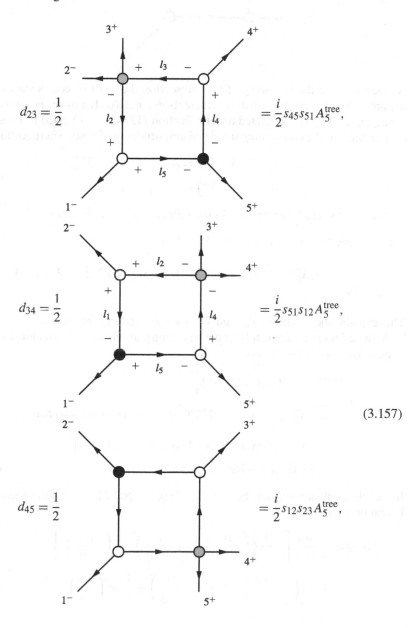

$$(3.157)$$

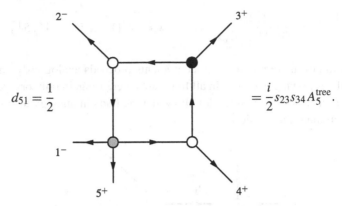

$$d_{51} = \frac{1}{2} \qquad\qquad = \frac{i}{2} s_{23} s_{34} A_5^{\text{tree}}.$$

As mentioned at the beginning of this subsection the last two coefficients d_{45} and d_{51} are related by reflection to the coefficients d_{23} and d_{34}. Let us see how this works for d_{45}. Clearly $I(s_{45})$ is related under reflection $(12345) \to (21543)$ to $I(s_{34})$. It is easy to see that the constituting tree-level amplitudes are related via reflection by

$$A_4\left(1^-,2^-,3^+,4^+\right) = A_4\left(2^-,1^-,4^+,3^+\right),$$
$$A_3\left(1^-,2^-,3^+\right) = -A_3\left(2^-,1^-,3^+\right), \tag{3.158}$$

leading to an overall sign change for the reflected on-shell diagram

$$I_{45} = \text{Reflect}[I_{34}] = -I_{34}\left(2^-,1^-,5^+,4^+,3^+\right)$$
$$= -\frac{i}{2} s_{12} s_{32} A_5^{\text{tree}}\left(2^-,1^-,5^+,4^+,3^+\right) = \frac{i}{2} s_{12} s_{23} A_5^{\text{tree}}\left(1^-,2^-,3^+,4^+,5^+\right). \tag{3.159}$$

This explains the results for d_{45} and d_{51} once d_{23} and d_{34} are proven.

Summarizing the final result for the box-contributions to the five-gluon amplitude in pure Yang-Mills theory reads

$$A_5^{\text{1-loop}}\left(1^-,2^-,3^+,4^+,5^+\right)$$
$$= \frac{i}{2} A_5^{\text{tree}}\left(1^-,2^-,3^+,4^+,5^+\right)\left[s_{34} s_{45} I(s_{12}) + s_{45} s_{51} I(s_{23})\right.$$
$$\left. + s_{51} s_{12} I(s_{34}) + s_{12} s_{23} I(s_{45}) + s_{23} s_{34} I(s_{51})\right]$$
$$+ \text{triangles} + \text{bubbles} + \mathscr{R}. \tag{3.160}$$

In fact the explicit result for the one-mass box integral $I(s_{12})$ in dimensional regularization is [4]

$$I(s_{12}) = \frac{-2ic_\Gamma}{s_{34} s_{45}}\left[-\frac{1}{\varepsilon^2}\left\{\left(\frac{\mu}{-s_{34}}\right)^\varepsilon + \left(\frac{\mu}{-s_{45}}\right)^\varepsilon - \left(\frac{\mu}{-s_{12}}\right)^\varepsilon\right\}\right.$$
$$\left. + \text{Li}_2\left(1 - \frac{s_{12}}{s_{34}}\right) + \text{Li}_2\left(1 - \frac{s_{12}}{s_{45}}\right) + \frac{1}{2}\ln^2\left(\frac{s_{34}}{s_{45}}\right) + \frac{\pi^2}{6}\right], \tag{3.161}$$

with $c_\Gamma = \frac{\Gamma(1+\varepsilon)\Gamma^2(1-\varepsilon)}{(4\pi)^{2-\varepsilon}\Gamma(1-2\varepsilon)}$. Plugging this into Eq. (3.160) we see that all the Mandelstam s_{ij} prefactors cancel out and we end up with a rather nice and compact result.

Exercise 3.1 (The Remaining Box Coefficients of $A_5^{\text{1-loop}}(1^-,2^-,3^+,4^+,5^+)$)
    Complete the discussion above and derive the remaining coefficients d_{23} and d_{34} as quoted in Eq. (3.157) using the same methods.

3.5.3 Rational Part: All-Plus Helicity Amplitudes

The reader might think that we have been overly careful in keeping terms that arise from the mismatch between the external four-dimensional momenta and the internal $(4-2\varepsilon)$-dimensional ones. Here we would like to illustrate that these terms are important and have to be kept in order to obtain the correct answer. This is true in QCD, as well as in supersymmetric gauge theories at higher loops, at least in the conventionally used schemes.
    The best example for the emergence of a rational term is perhaps the four-gluon all-plus helicity amplitude in pure Yang-Mills theory. It vanishes at tree-level, thanks to its hidden supersymmetry, as we have seen in Sect. 2.7.3. Renormalizability then requires that its one-loop contribution should be finite, as there is no possible counter term (given that the tree-level amplitude is zero). Therefore we can expect the result to be relatively simple. As we will see, it is in fact purely given by a rational term.
    Due to its simplicity the calculation may be performed e.g. using Feynman diagrams. We will quote the result here and then justify it by a calculation based on unitarity. One has

$$A_{\text{one-loop}}\left(1^+,2^+,3^+,4^+\right) = \frac{[12][34]}{\langle12\rangle\langle34\rangle}\frac{i}{(4\pi)^{D/2}}K_4, \qquad (3.162)$$

where

$$K_4 = \int \frac{d^4k\,d^{-2\varepsilon}\mu}{i\pi^{D/2}}$$

$$\times \frac{\mu^4}{(k^2-\mu^2)((k+p_1)^2-\mu^2)((k+p_1+p_2)^2-\mu^2)((k-p_4)^2-\mu^2)}. \qquad (3.163)$$

Note that in the nomenclature of Sect. 3.4 we would have written $\mu = l \cdot n_\varepsilon$. One can show that this integral evaluates to $-1/6$, despite the fact that its integrand naively vanishes in four dimensions. Therefore we have

$$A_{\text{one-loop}}\left(1^+,2^+,3^+,4^+\right) = -\frac{1}{6}\frac{i}{(4\pi)^{D/2}}\frac{[12][34]}{\langle12\rangle\langle34\rangle}. \qquad (3.164)$$

Fig. 3.9 Example caption

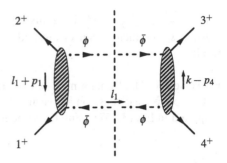

We would now like to reproduce the result for the D-dimensional integrand of the one-loop four-gluon amplitude $A(1^+, 2^+, 3^+, 4^+)$, cf. Eq. (3.162), from D-dimensional unitarity. We will follow closely references [11, 12]. In principle we have to compute this amplitude with gluons running in the loop. However, we can use the following argument to rewrite the pure-gluon amplitude in terms of supersymmetric and scalar gauge theories: in $\mathscr{N} = 4$ super Yang-Mills theory the one-loop n-gluon amplitude follows from summing the contributions of one gluon, four Weyl fermions and three complex (or six real) scalars running in the loop[5]

$$A^{\mathscr{N}=4}_{n,\text{one-loop}} = A^{[1]}_{n,\text{one-loop}} + 4 A^{[1/2]}_{n,\text{one-loop}} + 3 A^{[0]}_{n,\text{one-loop}}. \qquad (3.165)$$

Similarly for an n-gluon amplitude in $\mathscr{N} = 1$ super Yang-Mills theory containing one gluon and one gluino we have

$$A^{\mathscr{N}=1}_{n,\text{one-loop}} = A^{[1]}_{n,\text{one-loop}} + A^{[1/2]}_{n,\text{one-loop}}. \qquad (3.166)$$

Let us now specialize these two relations to the case of a four-gluon all-plus helicity amplitude. As follows from the discussion in Sect. 2.7.3 the amplitudes in the supersymmetric gauge theories have to vanish for this helicity configuration, $A^{\mathscr{N}=1 \text{ or } 4}_{n,\text{one-loop}}(1^+, 2^+, 3^+, 4^+) = 0$. This immediately implies the useful relation

$$A^{[1]}_{\text{one-loop}}\left(1^+, 2^+, 3^+, 4^+\right) = -A^{[1/2]}_{\text{one-loop}}\left(1^+, 2^+, 3^+, 4^+\right)$$

$$= A^{[0]}_{\text{one-loop}}\left(1^+, 2^+, 3^+, 4^+\right). \qquad (3.167)$$

Hence, in order to find the pure Yang-Mills amplitude $A^{[1]}_{\text{one-loop}}(1^+, 2^+, 3^+, 4^+)$, discussed above, we only need to consider an amplitude where a complex scalar runs in the loop.

Let us compute this amplitude using unitarity cuts. In order to evaluate the two-particle cuts, we will need a four-point tree-level amplitude of two gluons scattering into two scalars. Note that the scalars need to be kept in D dimensions, whereas the external states (gluons) are in four dimensions. Note that this is equivalent to

---

[5]With the upper indices $[i]$ we denote the spins of the respective particles running in the loop.

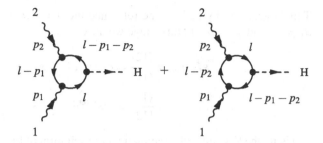

Fig. 3.10 Higgs boson production via fusion of two helicity plus gluons via a massive quark loop

computing a four-dimensional amplitude with massive scalars (however, in the one-loop integral, we will integrate over the mass of the scalars with a certain measure). This amplitude was computed in Sect. 2.5 as an example on the massive on-shell recursion. One has, cf. Eq. (2.83)

$$A\left(l_1^s, 1^+, 2^+, l_2^s\right) = \mu^2 \frac{[12]}{\langle 12 \rangle [(l_1 + p_1)^2 - \mu^2]}, \qquad (3.168)$$

where we have explicitly separated the D-dimensional and four-dimensional part of the loop momentum. Sewing this tree-level amplitude into the two-particle unitarity cut shown in Fig. 3.9, we see that it is in perfect agreement with Eq. (3.162).

3.5.4 An Example: Higgs Production via Gluon Fusion

Let us finish our discussion of generalized unitarity techniques with a calculation relevant for the dominant production channel of the Higgs boson at the Large Hadron Collider. It is given by the fusion of two gluons into a Higgs boson via the coupling to a top quark loop, see Fig. 3.10. This example was also presented in detail in Ref. [13]. For the time being we keep the gluon polarizations general. The amplitude including color structure takes the form

$$\mathcal{M}_{h_1 h_2}^{a_1 a_2} = \delta_{a_1 a_2} g_H \alpha_S \mathscr{A}_{h_1 h_2}, \qquad \mathscr{A}_{h_1 h_2} = \int \frac{d^D l}{(2\pi)^D} A_{h_1 h_2}(p_1, p_2, l), \qquad (3.169)$$

where g_H is the Higgs-Top Yukawa coupling constant and α_S denotes the strong coupling constant. We note the kinematical relations

$$p_1^2 = 0 = p_2^2, \qquad (p_1 + p_2)^2 = m_H^2, \qquad (3.170)$$

with the Higgs mass m_H. The fermion mass in the loop is denoted by m. We choose a gauge for the two gluon polarization vectors $\varepsilon_{1/2}^\mu$ leading to

$$p_i \cdot \varepsilon_j = 0 \quad \forall i, j = 1, 2. \qquad (3.171)$$

This is achieved by picking the reference momenta $\mu_i^\alpha \tilde{\mu}_i^{\dot\alpha} = q_i^{\alpha\dot\alpha}$ in Eq. (1.82) as $q_1 = p_2$ and $q_2 = p_1$. In this gauge we have

$$\varepsilon_1^+ \cdot \varepsilon_2^+ = \frac{[12]}{\langle 12 \rangle} = -e^{i\phi_{12}}, \qquad \varepsilon_1^+ \cdot \varepsilon_2^- = 0,$$

$$\varepsilon_1^- \cdot \varepsilon_2^- = \frac{\langle 12 \rangle}{[12]} = -e^{-i\phi_{12}}, \qquad \varepsilon_1^- \cdot \varepsilon_2^+ = 0. \tag{3.172}$$

Using the Feynman diagrammatic representation of Fig. 3.10 the integrand of the one-loop amplitude then reads using the massive fermion propagators $\frac{\slashed{l}+m}{l^2-m^2}$

$$A_{h_1 h_2} = \frac{\mathrm{Tr}[(\slashed{l}+m)\slashed{\varepsilon}_1(\slashed{l}-\slashed{p}_1+m)\slashed{\varepsilon}_2(\slashed{l}-\slashed{p}_1-\slashed{p}_2+m)]}{d_0 d_1 d_2} + (1 \leftrightarrow 2), \tag{3.173}$$

with

$$d_0 = l^2 - m^2, \qquad d_1 = (l-p_1)^2 - m^2,$$

$$d_2 = (l-p_2)^2 - m^2, \qquad d_3 = (l-p_1-p_2)^2 - m^2. \tag{3.174}$$

Note that the contribution of the second diagram in Fig. 3.10 is captured by the swap of labels $(1 \leftrightarrow 2)$. By evaluating the trace in the numerator it is clear that only terms proportional to m or m^3 can contribute, as the trace of an odd number of Dirac matrices vanishes. Using the well known trace identities of Eq. (B.7) we find

$$\mathrm{Tr}[\cdots] = m\,\mathrm{Tr}\big[\slashed{\varepsilon}_1(\slashed{l}-\slashed{p}_1)\slashed{\varepsilon}_2(\slashed{l}-\slashed{p}_1-\slashed{p}_2)\big] + m\,\mathrm{Tr}\big[\slashed{l}\slashed{\varepsilon}_1\slashed{\varepsilon}_2(\slashed{l}-\slashed{p}_1-\slashed{p}_2)\big]$$

$$+ m\,\mathrm{Tr}\big[\slashed{l}\slashed{\varepsilon}_1(\slashed{l}-\slashed{p}_1)\slashed{\varepsilon}_2\big] + m^3\,\mathrm{Tr}[\slashed{\varepsilon}_1\slashed{\varepsilon}_2]$$

$$= 4m\left[4(\varepsilon_1 \cdot l)(\varepsilon_2 \cdot l) - \frac{m_H^2}{2}(\varepsilon_1 \cdot \varepsilon_2) - (\varepsilon_1 \cdot \varepsilon_2)d_1\right]. \tag{3.175}$$

Hence the integrand takes the rather compact form

$$A_{h_1 h_2} = 4m\left[\frac{4(\varepsilon_1 \cdot l)(\varepsilon_2 \cdot l) - \frac{m_H^2}{2}(\varepsilon_1 \cdot \varepsilon_2)}{d_0 d_1 d_3} - \frac{(\varepsilon_1 \cdot \varepsilon_2)}{d_0 d_3} + (1 \leftrightarrow 2)\right]. \tag{3.176}$$

It will decompose into triangle, bubble and tadpole integrands under reduction as

$$A_{h_1 h_2} = \frac{c_1(l)}{d_0 d_1 d_3} + \frac{c_2(l)}{d_0 d_2 d_3} + \frac{b_{03}(l)}{d_0 d_3} + \cdots. \tag{3.177}$$

Where the dots stand for bubbles and tadpoles with massless leg basis integral depending only on the fermion mass m.

There is an important observation to make. The considered process is the leading order, there is no tree-level contribution due to the absence of a $(gg\phi)$-interaction vertex in the theory. This fact and the renormalizability of the standard model Lagrangian implies that the one-loop process under consideration must be UV finite simply because there is no possible counterterm to remove a UV divergence. As the bubble and tadpole integrals have UV divergences they cannot contribute to the final result. To be precise the bubbles could only contribute to a rational term, as that is finite. There can be no contribution from tadpoles.

Following the general strategy laid out above we need to find the triple cut loop momentum obeying the on-shell conditions $d_0 = d_1 = d_3 = 0$ or

$$l^2 = m^2, \qquad 2l \cdot p_1 = 0, \qquad 2l \cdot p_2 = m_H^2. \qquad (3.178)$$

The two one parameter solutions of these conditions may be constructed as[6]

$$l_\pm^\mu = p_1^\mu \pm \sqrt{\frac{1}{2\varepsilon_1 \cdot \varepsilon_2}(m^2 - \mu^2)}\left(c\varepsilon_1^\mu + \frac{1}{c}\varepsilon_2^\mu\right) + \mu n_\varepsilon^\mu, \quad \mu := l \cdot n_\varepsilon, \qquad (3.179)$$

with an arbitrary parameter $z \in \mathbb{C}$. Plugging either solution into the numerator of the triangle term in Eq. (3.176) yields the same expression

$$4(\varepsilon_1 \cdot l_\pm)(\varepsilon_2 \cdot l_\pm) = 2(m^2 - \mu^2)(\varepsilon_1 \cdot \varepsilon_2). \qquad (3.180)$$

Hence we have shown that

$$c_1(l) = \lim_{d_i \to 0} d_0 d_1 d_3 A_{h_1 h_2} = 4m(\varepsilon_1 \cdot \varepsilon_2)\left[2m^2 - \frac{1}{2}m_H^2 - 2\mu^2\right]. \qquad (3.181)$$

As the result is symmetric under $(1 \leftrightarrow 2)$ we conclude that the coefficient of the second triple cut is identical, i.e. $c_2(l) = c_1(l)$.

We proceed with the determination of the bubble coefficient $b_{03}(l)$ of Eq. (3.177) by considering the double cut $d_0 = d_3 = 0$, i.e.

$$l^2 = m^2, \qquad 2l \cdot (p_1 + p_2) = m_H^2. \qquad (3.182)$$

The double cut loop momentum may be parametrized as

$$l^\mu = \frac{1}{2}(p_1 + p_2)^\mu + \left[ap_{12}^\mu + b\varepsilon_1^\mu + c\varepsilon_2^\mu\right] + \mu n_\varepsilon^\mu. \qquad (3.183)$$

Note that $(p_1 + p_2)$, $p_{12} = p_1 - p_2$, $(b\varepsilon_1 + c\varepsilon_2)$ and n_ε are all orthogonal to each other. The three parameters a, b, c above are subject to the condition

$$2bc(\varepsilon_1 \cdot \varepsilon_2) = m^2 - \mu^2 - m_H^2\left(\frac{1}{4} - a^2\right) \qquad (3.184)$$

in order to satisfy Eq. (3.182). The bubble coefficient follows from

$$b_{03}(l) = \lim_{d_0, d_3 \to 0} d_0 d_3 \left[A_{h_1 h_2} - \frac{c_1(l)}{d_0 d_1 d_3} - \frac{c_2(l)}{d_0 d_2 d_3}\right], \qquad (3.185)$$

---

[6]Recall that $\varepsilon_i \cdot \varepsilon_i = 0$ and that in our gauge choice $p_i \cdot \varepsilon_j = 0$.

with $c_1(l) = c_2(l)$ just as given in Eq. (3.181), as the l dependence is only through the $\mu = l \cdot n_\varepsilon$ being identical in Eqs. (3.179) and (3.183). We now compute for the double cut loop momentum (3.183)

$$d_1 = -2l \cdot p_1 = m_H^2 \left(a - \frac{1}{2} \right),$$

$$d_2 = -2l \cdot p_2 = -m_H^2 \left(a + \frac{1}{2} \right), \qquad (3.186)$$

$$2(\varepsilon_1 \cdot \varepsilon_2) = (\varepsilon_1 \cdot \varepsilon_2) \left[m^2 - \mu^2 + m_H^2 \left(a^2 - \frac{1}{4} \right) \right].$$

Armed with these one has

$$\lim_{d_0, d_3 \to 0} d_0 d_3 A_{h_1 h_2}$$

$$= 4m(\varepsilon_1 \cdot \varepsilon_2) \left[2 \left[m^2 - \mu^2 + m_H^2 \left(a^2 - \frac{1}{4} \right) - \frac{m_H^2}{2} \right] \left(\frac{1}{d_1} + \frac{1}{d_2} \right) - 2 \right],$$

$$(3.187)$$

$$\lim_{d_0, d_3 \to 0} c_1(l) \left(\frac{1}{d_1} + \frac{1}{d_2} \right) = 4m(\varepsilon_1 \cdot \varepsilon_2) \left[2(m^2 - \mu^2) - \frac{1}{2} m_H^2 \right] \left(\frac{1}{d_1} + \frac{1}{d_2} \right).$$

$$(3.188)$$

The difference of these two relations yields the bubble coefficient $b_{01}(l)$ which nicely vanishes

$$b_{03}(l) = 4m(\varepsilon_1 \cdot \varepsilon_2) \left[2m_H^2 \left(a^2 - \frac{1}{4} \right) \left(\frac{1}{d_1} + \frac{1}{d_2} \right) - 2 \right] = 0. \qquad (3.189)$$

Hence there is no contribution to the rational piece from this double cut channel. The same holds true for the other double cuts, which we refrain from discussing here.

We are now in the position to state the final result for the amplitude. The last observation to be made is, that the three-point scalar and rational part integrals emerging from the first two terms in the integrated version of Eq. (3.177) are identical, as they can be related to each other through a shift $p_1 + p_2$ in the loop momentum l. Putting everything together we state the final form of the Higgs production via gluon fusion amplitude

$$\mathscr{A}_{h_1 h_2} = 8m(\varepsilon_1 \cdot \varepsilon_2) \left[\left(2m^2 - \frac{1}{2} m_H^2 \right) I_3 - 2 I_3 (\mu^2) \right]. \qquad (3.190)$$

Here I_3 is three-point scalar integral $I_3(0, 0, m_H^2; m^2, m^2, m^2)$ in the nomenclature of Eq. (3.18) and $I_3(\mu^2)$ contributes to the rational piece of the amplitude

$$-i(4\pi)^{D/2}I_3 = \int \frac{d^D l}{i\pi^{D/2}} \frac{1}{d_0 d_1 d_3} = \frac{2}{m_H^2} f\left(\frac{4m^2}{m_H^2}\right)$$

$$\text{where } f(\tau) = \begin{cases} -\arcsin^2(\tau^{-1/2}) & \tau > 1, \\ \frac{1}{4}(\log \frac{1+\sqrt{1-\tau}}{1-\sqrt{1-\tau}} - i\pi)^2 & \tau \le 1, \end{cases} \tag{3.191}$$

$$-i(4\pi)^{D/2}I_3(\mu^2) = \int \frac{d^D l}{i\pi^{D/2}} \frac{(l \cdot n_\varepsilon)^2}{d_0 d_1 d_3} = -\frac{1}{2}.$$

Techniques for the evaluation of these three-point integrals will be discussed in Sect. 3.8, in particular in Exercise 3.3 and Sect. 3.8.3 where the analytic result of I_3 will be derived.

Finally we note that the amplitude Eq. (3.190) vanishes if the helicities of the gluons are opposite

$$\mathscr{A}_{+-} = 0 = \mathscr{A}_{-+}. \tag{3.192}$$

This may be understood physically in the Higgs boson rest frame: the Higgs boson is a spinless particle produced at rest by the two gluons coming in with opposite three-momenta $\pm \mathbf{q}$. Hence the helicities of the fusing gluons must be identical by conservation of spin. The final result reads

$$|\mathscr{A}_{++}| = |\mathscr{A}_{--}| = \frac{m}{2\pi^2}|(\tau - 1)f(\tau) + 1|, \qquad \tau = \frac{4m^2}{m_H^2}. \tag{3.193}$$

3.6 Overview of Integration Techniques for Loop Integrals

In the remaining sections of this chapter we turn to the discussion of a number of integration techniques for the evaluation of Feynman loop integrals. After setting up our conventions for integrals in D-dimensional Minkowski space, we describe the Feynman/Schwinger parameter technique, which allows to evaluate simple loop integrals. It will also provide the starting point for switching to Mellin-Barnes representations. In later sections, we discuss the integration by parts identities, and explain how they can be used in order to derive differential equations for loop integrals.

We are following the conventions of Refs. [14, 15]. This reference should also be consulted for further reading. Several advanced topics will not be covered in these introductory notes. We give further references to the literature at the end of this section. In general the aim of the present discussion is to provide the reader with the main ideas, and to give the most important formulas that are relevant in practice. Examples are mostly delegated to the exercises, so that the main formulas are easily found.

Fig. 3.11 Generic one-loop integral in momentum and dual representation

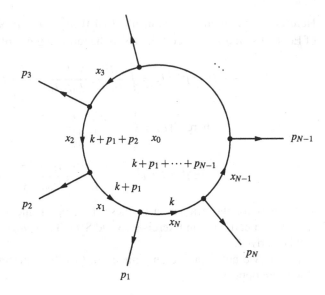

3.6.1 Notation and Dual Coordinates

Let us start with a generic one-loop scalar n-point integral departing slightly from the former conventions of Eq. (3.12)

$$F_n = \int \frac{d^D k}{i\pi^{D/2}}$$

$$\times \frac{1}{(-k^2 + m_1^2)^{a_1}(-(k+p_1)^2 + m_2^2)^{a_2}\cdots(-(k+p_1+\cdots+p_{n-1})^2 + m_n^2)^{a_n}},$$
(3.194)

see Fig. 3.11, where the external momenta p_i may or may not satisfy on-shell conditions. It is convenient to introduce dual, or region coordinates,

$$k = x_0 - x_1, \qquad p_j = x_j - x_{j+1},$$
(3.195)

with the identification $x_{j+N} \equiv x_j$. Note that the relation to the region momenta q_i introduced previously in Eq. (3.6) reads $x_j = -q_{j-1}$. Then the integral above takes the simple form

$$F_n = \int \frac{d^D x_0}{i\pi^{D/2}} \prod_{j=1}^{n} \frac{1}{(-x_{0j}^2 + m_j^2)^{a_j}},$$
(3.196)

where $x_{ij} = x_i - x_j$. Translation invariance in the dual space corresponds to the freedom of redefining the loop integration variables k in the initial integral. Momentum conservation implies that the momenta form a closed polygon in dual space, with

the vertices being the x_i and the edges being the p_i. Beyond one loop, dual coordinates can only consistently be introduced for planar integrals, but the above
formulas stay useful for one-loop subintegrals. This can be useful when switching
to Mellin-Barnes parametrizations, see Sect. 3.7, which can be done one loop at a
time.

3.6.2 Feynman Parametrization

The idea is to relate the generic one-loop integral as given in Eq. (3.196) to the
integral over one propagator, cf. Eq. (3.4), at the cost of introducing auxiliary integrations. The space-time integration can then be carried out, and one is left with an
integral over the auxiliary integration variables.

There are two closely related tricks for achieving this, Feynman and Schwinger
parametrizations. We have already encountered the latter in Eq. (3.3). The former is
based on the identity

$$\frac{1}{X_1^{a_1} X_2^{a_2}} = \frac{\Gamma(a_1 + a_2)}{\Gamma(a_1)\Gamma(a_2)} \int_0^1 d\alpha_1 d\alpha_2 \frac{\delta(1 - \alpha_1 - \alpha_2)\alpha_1^{a_1-1}\alpha_2^{a_2-1}}{(\alpha_1 X_1 + \alpha_2 X_2)^{a_1+a_2}}. \tag{3.197}$$

One can easily prove the generalization to n propagators by induction,

$$\frac{1}{\prod_{i=1}^n X_i^{a_i}} = \frac{\Gamma(a_1 + \cdots + a_n)}{\prod_{i=1}^n \Gamma(a_i)} \int_0^1 \left[\prod_{i=1}^n d\alpha_i \alpha_i^{a_i-1} \right] \frac{\delta(1 - \sum_{i=1}^n \alpha_i)}{(\sum_{i=1}^n \alpha_i X_i)^{a_1+\cdots+a_n}}. \tag{3.198}$$

Applying Eq. (3.198) to the one-loop integral (3.196) yields an integral over a simple propagator. The latter integral can be carried out by performing a change of
variables, and using Eq. (3.4)

$$F_n = \frac{\Gamma(a - D/2)}{\prod_{i=1}^n \Gamma(a_i)} \times \int_0^\infty \left[\prod_{i=1}^n d\alpha_i \alpha_i^{a_i-1} \right] \frac{\delta(\sum_{i=1}^n c_i \alpha_i - 1) U^{a-D}}{(V + U \sum_{i=1}^n m_i^2 \alpha_i)^{a-D/2}}. \tag{3.199}$$

Here $U = \sum_{i=1}^n \alpha_i$ and $V = \sum_{i<j} \alpha_i \alpha_j (-x_{ij}^2)$. They, as well as their higher-loop
analogs, have graph theoretical interpretations, see e.g. [14, 15].

In fact, using the formulas given above one arrives at Eq. (3.199) with all $c_i = 1$
in the delta function. One can show that the integral above is valid for any values of
the c_i, as long as one of them is non-zero. There exists a generalization of the above
formula for arbitrary loop order, cf. [14, 15], but we will only need the one-loop
case in the following.

As an example, let us consider the massless on-shell triangle diagram of
Fig. 3.12,

$$F_3 = \int \frac{d^D k}{i \pi^{D/2}} \frac{1}{k^2 (k + p_1)^2 (k - p_2)^2}, \tag{3.200}$$

Fig. 3.12 Massless triangle
diagram discussed in the
main text

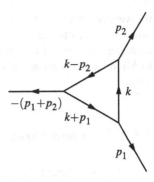

with on-shell conditions $p_1^2 = p_2^2 = 0$. The integral depends on the dimensionful
variable $s = (p_1 + p_2)^2$, and on ε. Here we have dropped the explicit $+i0$ prescription. In the region $-s > 0$, the integral is real, and for other regions one can recover
the $+i0$ prescription by giving a small imaginary part to s, $s \to s + i0$. This integral has infrared divergences coming from soft and collinear regions of the loop
integration, which together produce a $1/\varepsilon^2$ pole, as we will see.

Our main one-loop formula Eq. (3.199) immediately gives (with $c_i = 1$)

$$F_3 = \Gamma(1+\varepsilon) \int d\alpha_1 d\alpha_2 d\alpha_3 \frac{\delta(1 - \alpha_1 - \alpha_2 - \alpha_3)}{(-s\alpha_1\alpha_3)^{1+\varepsilon}}. \tag{3.201}$$

The dependence on s is trivial and is dictated by the overall dimensionality of the integral. Note that this expression is valid for $\varepsilon < 0$, since we are dealing with infrared
divergences. We carry out the α integrals using

$$\int \prod_{i=1}^{n} d\alpha_i \alpha_i^{a_i-1} \delta\left(\sum_{i=1}^{n} \alpha_i - 1\right) = \frac{\prod_{i=1}^{n} \Gamma(a_i)}{\Gamma(\sum_{i=1}^{n} \alpha_i)}, \tag{3.202}$$

and find

$$F_3 = (-s)^{-1-\varepsilon} \frac{\Gamma(1+\varepsilon)\Gamma^2(-\varepsilon)}{\Gamma(1 - 2\varepsilon)}. \tag{3.203}$$

We wish to expand this formula for small ε. The Γ function has the expansion

$$\log(\Gamma(1+\varepsilon)) = -\gamma_E\varepsilon + \sum_{n=2}^{\infty} (-1)^n \zeta_n \frac{\varepsilon^n}{n}, \tag{3.204}$$

where the zeta values are defined as

$$\zeta_n = \sum_{k=1}^{\infty} \frac{1}{k^n}, \tag{3.205}$$

and $\gamma_E = 0.57721\ldots$ is the Euler-Mascheroni constant. Using this formula, and $\Gamma(1+n) = n\Gamma(n)$, we find

$$e^{\varepsilon\gamma_E} F_3 = (-s)^{-1-\varepsilon}\left[\frac{1}{\varepsilon^2} - \frac{\pi^2}{12} - \frac{7}{3}\zeta_3\varepsilon - \frac{47\pi^4}{1440}\varepsilon^2 + \mathcal{O}(\varepsilon^3)\right]. \tag{3.206}$$

Here we multiplied F_3 by a factor of $e^{\varepsilon\gamma_E}$ (in general, one takes one such factor per loop order), in order to avoid the explicit appearance of γ_E in the expansions.

3.7 Mellin-Barnes Techniques

In the previous section we saw how to evaluate simple Feynman integrals using Feynman parametrization. In some cases, it may be advantageous not to evaluate the Feynman parameter integrals directly, but to switch to another representation. The idea is to trade the Feynman parameter integrals for Mellin-Barnes integrals, as we describe presently.

The key formula is the following,

$$\frac{1}{(x+y)^a} = \frac{1}{\Gamma(a)} \int_{c-i\infty}^{c+i\infty} \frac{dz}{2\pi i}\Gamma(-z)\Gamma(z+a)x^z y^{-a-z}, \tag{3.207}$$

where the integration contour is parallel to the imaginary axis, with real part c in the interval $-a < c < 0$. See Fig. 3.13. In general, the integration contour is chosen such that the poles of Γ functions of the type $\Gamma(z + \cdots)$ lie to its left, and the poles of $\Gamma(-z + \cdots)$ lie to its right.

One can verify the validity of Eq. (3.207) by thinking about a series expansion of its left hand side. Assume $x < y$. Then, we have

$$\frac{1}{(x+y)^a} = y^{-a}\sum_{n\geq 0}(-1)^{n+1}\frac{\Gamma(n+a)}{\Gamma(n+1)\Gamma(a)}\left(\frac{x}{y}\right)^n, \quad x < y. \tag{3.208}$$

On the other hand, if $x < y$ we can close the integration contour in Eq. (3.207) on the right, because the contribution from the semicircle at infinity vanishes. By complex analysis, we get a contribution from (minus) all poles of $\Gamma(-z)$ situated at $z_n = n$ ($n = 0, 1, \ldots$). Summing up the corresponding residues, $\text{Res}[\Gamma(-n)] = (-1)^{n+1}/n!$, one finds perfect agreement with Eq. (3.208). One may verify similarly the validity of Eq. (3.207) for $y < x$. In this case, one closes the integration contour on the left.

Equation (3.207) can be used to factorize expressions, e.g. the denominator factors appearing in Feynman parametrization. Once factorized, Eq. (3.202) allows to carry out the Feynman parameter integrals. In some sense, the Mellin-Barnes representation can therefore be considered the inverse of the Feynman parametrization. Of course, this means that one is just trading one kind of integral representation

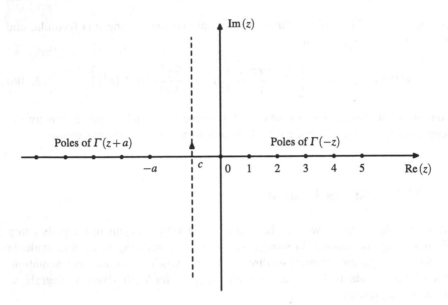

Fig. 3.13 Integration contour in the Mellin-Barnes representation considered in the main text

for another. However, the Mellin-Barnes representation is very flexible, and has a number of useful features, as we will see shortly.

Let us apply the above procedure to the massless one-loop box integral, cf. Fig. 3.14

$$F_4 = \int \frac{d^D k}{i\pi^{D/2}} \frac{1}{k^2(k+p_1)^2(k+p_1+p_2)^2(k+p_1+p_2+p_3)^2}. \qquad (3.209)$$

It is a function of $s = (p_1+p_2)^2$, $t = (p_2+p_3)^2$, and ε. As explained in the previous section, we can write down a Feynman parametrization using Eq. (3.199),

$$F_4 = \Gamma(a - D/2) \int [d\alpha_4] \frac{1}{[\alpha_1\alpha_3(-s) + \alpha_2\alpha_4(-t)]^{a-D/2}}. \qquad (3.210)$$

Here $a = 4$ and $D = 4 - 2\varepsilon$, and we introduced the shorthand notation

$$[d\alpha_n] := \prod_{i=1}^{n} d\alpha_i \delta\left(\sum_{i=1}^{n} \alpha_i - 1\right). \qquad (3.211)$$

We can factorize the integrand at the cost of introducing one Melin-Barnes parameter integral, using (3.207). Then, the integral over the α parameters can be done, cf. Eq. (3.202), and we find

$$F_4 = \int \frac{dz}{2\pi i} (-s)^z (-t)^{-2-\varepsilon-z} \Gamma(-z) \Gamma(2+\varepsilon+z) \frac{\Gamma^2(1+z)\Gamma^2(-1-\varepsilon-z)}{\Gamma(-2\varepsilon)}. \qquad (3.212)$$

Fig. 3.14 Massless
four-point function
considered in the main text

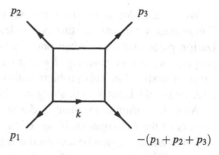

For the integrations leading to this expression to be well defined, the real part of the arguments of each Γ function must be positive. We see that this implies in particular that $\varepsilon < 0$, which is expected since the integral is IR divergent. We can choose e.g.

$$\mathbf{Re}(z) = -1/2, \qquad \varepsilon = -1. \tag{3.213}$$

We will now explain how to analytically continue to $\varepsilon \to 0$.

3.7.1 Resolution of Singularities in ε

We saw that the integral in Eq. (3.212) cannot be defined for $\varepsilon = 0$. This can be traced back to the presence of the Gamma functions $\Gamma(1+z)$ and $\Gamma(-1-\varepsilon-z)$. The contour for the z integration has to pass between the poles of these Gamma functions, which is only possible for $\varepsilon < 0$. Before we can take the limit $\varepsilon \to 0$, we must therefore deform the integration contour for z, e.g. to the right, which leads to a contribution of the residue at $z = -1 - \varepsilon$. The value of this residue is

$$(-t)^{-1-\varepsilon}(-s)^{-1}\frac{\Gamma^2(-\varepsilon)\Gamma(\varepsilon+1)}{\Gamma(-2\varepsilon)}\left(\frac{s}{t}\right)^{-\varepsilon}$$

$$\times \left[2\psi^{(0)}(-\varepsilon) - \psi^{(0)}(\varepsilon+1) + \log\frac{s}{t} + \gamma_{\mathrm{E}}\right]. \tag{3.214}$$

Here $\psi^{(0)}(z) = \partial_z \log(\Gamma(z))$. In the remaining terms, we can safely set $\varepsilon = 0$. We see that they are of $\mathcal{O}(\varepsilon)$, due to the presence of the factor $\Gamma(-2\varepsilon)$ in the denominator. Therefore we find (remembering that the residue contributes with a minus sign),

$$e^{\varepsilon\gamma_{\mathrm{E}}}F = (-t)^{-1-\varepsilon}(-s)^{-1}\left[\frac{4}{\varepsilon^2} - \frac{2}{\varepsilon}\log\frac{s}{t} - \frac{4\pi^2}{3} + \mathcal{O}(\varepsilon)\right]. \tag{3.215}$$

The answer can be put into a physically more transparent form, which exposes the origin of the infrared divergences in the s and t-channel,

$$st e^{\varepsilon\gamma_{\mathrm{E}}}F = \frac{2}{\varepsilon^2}(-s)^{-\varepsilon} + \frac{2}{\varepsilon^2}(-t)^{-\varepsilon} - \log^2\frac{s}{t} - \frac{4\pi^2}{3}. \tag{3.216}$$

We would like to mention another case where the same technique is useful. Sometimes it can happen that one is lead to introduce auxiliary analytic regularization parameters, e.g. when one wants to treat a numerator factor as an inverse propagator. In this case, e.g. for exponent $a = -1$, the Feynman parametrization formula is singular. This problem can be avoided by first setting $a = -1 + \delta$, and later taking the limit $\delta \to 0$, along the above lines.

Also, it is often convenient to derive Mellin-Barnes representations for general powers of the propagators. If one of the propagators is absent, the limit is usually singular, and can again be analyzed using the above procedure.

Two different strategies of resolving singularities in Melin-Barnes integrals [16, 17] have been implemented in [18] and [19], respectively. An example is given at the end of this section.

3.7.2 Asymptotic Expansions and Resummation

We have seen above how the Mellin-Barnes approach allowed to obtain the full result for a one-loop box integral, in the limit $\varepsilon \to 0$. In more complicated cases, one can have multiple Mellin-Barnes integrals, and one or more integrals can remain after taking the limit $\varepsilon \to 0$. In those cases, one can try to close the contour of integration, which yields an infinite series representation of the result. One can then try to resum this expansion. In some cases, when the expansion depends on a parameter, this can be used efficiently to obtain an approximation where this parameter is small (or large). We show a simple example of this (another example can be found as an exercise), and refer the interested reader to the literature for more details.

Consider the following example,

$$f(x) := \int \frac{dz}{2\pi i} \Gamma^2(-z) \Gamma^2(1+z) x^{-1-z}, \quad -1 < \mathbf{Re}(z) < 0. \tag{3.217}$$

Let us imagine that $0 < x < 1$. By closing the contour on the left, we obtain an infinite series expansion, with contributions from residues at $z = -1, -2, \ldots$. The residue at $z = -n$, with $n \in \mathbf{N}$ is

$$-x^{-n+1} \log x, \tag{3.218}$$

and we deduce that

$$f(x) = -\frac{1}{1-x} \log x. \tag{3.219}$$

We also see that had we been interested only in the asymptotic expansion of $f(x)$ for $x \ll 1$, we could have kept only the first few residues. In particular, just keeping the first term at $z = -1$ gives $f(x) = -\log(x) + \mathcal{O}(x)$. Adding further residues systematically improves the approximation. This can be very useful in problems with several scales, where one has to take asymptotic limits. We will see this in an exercise where a mass parameter m^2 is small w.r.t. kinematical invariants $-s$, $-t$. Another important case is the Regge limit, where some kinematical invariants are taken to be large w.r.t. others. For summation of series in more complicated cases we refer the interested reader to [20–22].

3.7.3 Numerical Evaluation and Convergence

In cases where it is too difficult to obtain an analytical result, the Mellin-Barnes representation can also be very efficient to numerically evaluate the answer.

The convergence properties of these integrals are typically very good due to the presence of the Γ functions in the integrand, which have an exponential falloff in the complex plane. However, there are also factors of kinematical invariants, such as $(-s)^z$, as we have seen in examples. For Euclidean kinematics, i.e. $s < 0$, this factor just contributes a phase w.r.t. the integration along the imaginary axis. However, for $s > 0$ it becomes important and can in principle compensate the exponential falloff coming from the Γ functions. In these cases, a detailed study of the convergence properties is necessary. See Ref. [18] for a more detailed discussion.

Exercise 3.2 (Feynman and Melin-Barnes Parametrization)

Consider the one-loop box integral with uniform integral mass m^2, and massless external lines, $p_i^2 = 0$. This integral is UV and IR finite in four dimensions. It is given by

$$F_{4,m} = st \int \frac{d^4k}{i\pi^2}$$

$$\times \frac{1}{(-k^2 + m^2)(-(k+p_1)^2 + m^2)(-(k+p_1+p_2)^2 + m^2)(-(k-p_4)^2 + m^2)},$$
$$(3.220)$$

where momentum conservation implies $\sum_{i=1}^{4} p_i^\mu = 0$ and the on-shell conditions for massless external particles read $p_i^2 = 0$. We use the notation $s = (p_1 + p_2)^2$ and $t = (p_2 + p_3)^2$. Let us assume $s, t < 0$ for simplicity, such that $F_{4,m}$ will be real.

The parts (d) and (e) of this exercise refer to the very useful Mathematica implementation MB.m of Czakon [18]. All steps can also be done by hand but it is recommended to use the computer algebra.

(a) Quick question: How many variables does $F_{4,m}$ depend on? What are they?
(b) Derive a Feynman representation for $F_{4,m}$.
(c) Introduce Mellin-Barnes parameters to factorize the Feynman integrand and carry out the Feynman/alpha parameter integrations. Note that a tool for the automatic generation of Mellin-Barnes representations was published in [23].
(d) Find allowed values for the real parts of the Mellin-Barnes integrations. (You can also use MBrules of MB.m to do this.)
(e) Think about what happens when m goes to zero. Use MBaymptotics.m to perform the $m \to 0$ limit analytically.

3.8 Integration by Parts and Differential Equations

In the previous sections we discussed how to introduce Feynman and Mellin-Barnes representations for individual Feynman diagrams. In this section we discuss identities between different Feynman integrals. In a problem where many integrals need to be computed, this allows to reduce the calculation to a smaller set of master integrals. Knowing the reduction to master integrals also allows one to set up differential equations for the latter, which can be an efficient method for computing them.

3.8.1 Integration by Parts Identities

Integration by parts identities are derived by noticing that total derivatives in Feynman integrals vanish. Let us show the main idea using a simple example, and then generalize.

Consider the one-loop massive bubble integral, for arbitrary integer powers of the propagators,

$$J(a_1, a_2) := \int \frac{d^D k}{i\pi^{D/2}} \frac{1}{(-k^2 + m^2)^{a_1}(-(k+p)^2 + m^2)^{a_2}}. \tag{3.221}$$

The reason for considering integrals for arbitrary integer powers will become clear shortly: it will allow us to derive relations between different integrals. Indeed, we can derive an identity between integrals with different indices (a_1, a_2) by considering

$$0 = \int \frac{d^D k}{i\pi^{D/2}} \frac{\partial}{\partial k^\mu} \left(k^\mu \frac{1}{(-k^2 + m^2)^{a_1}(-(k+p)^2 + m^2)^{a_2}} \right). \tag{3.222}$$

In this way, after some algebra one obtains,

$$0 = (D - 2a_1 - a_2) J(a_1, a_2) - a_2 J(a_1 - 1, a_2 + 1)$$
$$+ 2m^2 a_1 J(a_1 + 1, a_2) + (2m^2 - p^2) a_2 J(a_1, a_2 + 1). \tag{3.223}$$

A similar identity follows from the symmetry $J(a_1, a_2) = J(a_2, a_1)$. This gives two equations for $J(a_1 + 1, a_2)$ and $J(a_1, a_2 + 1)$ in terms of integrals for which the sum of the indices equals $a_1 + a_2$. For $a_1 \neq 0, a_2 \neq 0$ this system is non-singular and therefore we can always express $J(b_1, b_2)$ in terms of integrals $J(a_1, a_2)$ with $a_1 + a_2 = b_1 + b_2 - 1$. Let us discuss also the special case if one of the initial indices is zero. For $a_2 = 0$, Eq. (3.223) simply becomes

$$0 = (D - 2a_1) J(a_1, 0) + 2m^2 a_1 J(a_1 + 1, 0), \tag{3.224}$$

and therefore any $J(a, 0)$ can be expressed in terms of $J(1, 0)$.

Therefore we see that any $J(a_1, a_2)$ for positive a_1, a_2 can be expressed in terms of $J(1, 0)$ and $J(1, 1)$ (and $J(0, 1) = J(1, 0)$ by symmetry). For example, we have

$$J(1, 2) = \frac{(D-2)}{2m^2(4m^2 - p^2)} J(1, 0) + \frac{(D-3)}{4m^2 - p^2} J(1, 1). \qquad (3.225)$$

We can choose $J(1, 0)$ and $J(1, 1)$ as master integrals of this family of integrals. Note that one can use the system of equations to define other master integrals. In the present case, we can choose any two integrals that form a basis. For example, as we will see later, in certain cases it may be advantageous to choose master integrals different from the 'canonical' choice. For example, one may define UV finite master integrals, such as $J(3, 0)$ and $J(2, 1)$, or, as we will see in Sect. 3.8.3, master integrals that obey simple differential equations.

These ideas for reducing integrals to master integrals can straightforwardly be generalized to arbitrary one-loop integrals with a higher number of external points. We refer the interested reader to [14, 15] for more details and references regarding the multi-loop case.

3.8.2 Differential Equations

The reduction to master integrals in the previous section can be used to set up (systems of) differential equations for the latter. The idea is to differentiate the master integrals with respect to external parameters, such as momenta or masses. This will in general give integrals of the same type, but with different powers of the propagators. The latter can then be reduced to master integrals using the reduction identities, giving a system of (first-order) differential equations.

Let us illustrate this in the previous example of bubble integrals. We choose $J(3, 0)$ and $J(2, 1)$ as master integrals, for $D = 4 - 2\varepsilon$. They are finite as $\varepsilon \to 0$.

We consider the elementary tadpole integral $J(3, 0)$ to be known, see Eq. (3.4),

$$J(3, 0) = \frac{\Gamma(3 - D/2)}{\Gamma(3)} \frac{1}{(m^2)^{3-D/2}}. \qquad (3.226)$$

Let us now determine $J(2, 1)$, which is a nontrivial function of m^2 and q^2. We can differentiate it w.r.t. m^2 and find the equation

$$\partial_{m^2} J(2, 1) = -2J(3, 1) - J(2, 2). \qquad (3.227)$$

Reducing the right-hand-side to master integrals works as follows. Taking the sum and the difference of Eq. (3.223) for $a_1 = 1, a_2 = 2$ and $a_2 = 1, a_2 = 1$ one has

$$0 = 2(D - 5)J(2, 1) - 2J(3, 0) + (4m^2 - p^2)\big[J(2, 2) + 2J(3, 1)\big], \qquad (3.228)$$

$$0 = 2J(2, 1) - 2J(3, 0) + p^2\big[J(2, 2) - 2J(3, 1)\big]. \qquad (3.229)$$

These equations allow us to rewrite the differential equation (3.227) as

$$\partial_{m^2} J(2,1) = \frac{2}{4m^2 - p^2}\left[(D-5)J(2,1) - J(3,0)\right]. \qquad (3.230)$$

This is a linear first-order differential equation for $J(2,1)$, with the inhomogeneous part known from Eq. (3.226). There is also a simple boundary condition, since $J(2,1)_{p^2=0} = J(3,0)$. This information, together with simple dimensional analysis, completely determines $J(2,1)$ as a functions of m^2, p^2 and ε.

One can solve the differential equation for $J(2,1)$ e.g. in a series expansion in ε. For simplicity, let us derive the solution for the finite term. On dimensional grounds, we have

$$J(2,1) = \frac{1}{p^2} f(y) + \mathcal{O}(\varepsilon), \quad y = -m^2/p^2. \qquad (3.231)$$

Expanding the differential equation to $\mathcal{O}(\varepsilon)$ accuracy, we find

$$yf'(y) + \frac{2y}{1+4y} f(y) = \frac{1}{1+4y}, \qquad (3.232)$$

together with the boundary condition $f(\infty) = 0$. The solution is, for $y > 0$,

$$f(y) = \frac{1}{\sqrt{1+4y}} \log\left(\frac{\sqrt{1+4y}-1}{\sqrt{1+4y}+1}\right). \qquad (3.233)$$

It can be analytically continued to other values of $y = -m^2/s$, taking into account the $i0$ prescription of the initial Feynman integral. One can remark that the discussion of the differential equation can be simplified by introducing the variable $x = (\sqrt{1+4y}-1)/(\sqrt{1+4y}+1)$ which is natural in this problem.

Exercise 3.3 (Integration by Parts and Differential Equations)

(a) Perturbative solution of differential equation in ε:

Write the master integral $J(2,1)$ considered in the main text in a series expansion in ε and study the structure of the differential equation (3.230). Compute the order ε and ε^2 terms of the expansion.

(b) Differential equation for triangle integral:

Consider the triangle integral

$$F_{3,m} = \int \frac{d^D k}{i\pi^{D/2}} \frac{1}{(-k^2+m^2)(-(k+p_1)^2+m^2))(-(k+p_2)^2+m^2)} \qquad (3.234)$$

and write down the integration by parts identities for this problem.

Next, consider $\partial_{m^2} F_{3,m}$ and use the integration by parts identities to simplify the inhomogeneous term in this equation. It is convenient to choose master integrals that are finite in four dimensions. Solve the differential equation for $F_{3,m}$ for $D=4$.

The solution is

$$F_{3,m}(D = 4) = -\frac{1}{2}\frac{1}{s}\log^2\left(\frac{\sqrt{1+4y}-1}{\sqrt{1+4y}+1}\right),\qquad(3.235)$$

where $y = -m^2/s$ and $s = (p_1 - p_2)^2$.

(c) Triangle identity:

Consider the massless two-loop propagators integral

$$Q(a_1, a_2, a_3, a_4, a_5; D)$$

$$:= \int \frac{d^D k_1 d^D k_2}{(i\pi^{D/2})^2}\frac{1}{(-k_1^2)^{a_1}(-(k_1+p)^2)^{a_2}}\frac{1}{(-(k_1-k_2)^2)^{a_3}(-(k_2+p)^2)^{a_4}(-k_2^2)^{a_5}},$$

$$(3.236)$$

and write down the integration by parts identities for one of the triangle sub integrals. Use the latter to express $Q(1, 1, 1, 1, 1; D)$ in terms of one-loop bubble integrals. Use the general formula for one-loop bubble integrals in order to express $Q(1, 1, 1, 1, 1, D)$ in terms of Γ functions. Show that

$$Q(1, 1, 1, 1, 1; 4 - 2\varepsilon) = \frac{1}{(-q^2)}6\zeta_3 + \mathcal{O}(\varepsilon).\qquad(3.237)$$

3.8.3 Simplified Approach to Differential Equations

Let us consider a family of one-loop triangle integrals with internal mass that already appeared in Exercise 3.3,

$$G(a_1, a_2, a_3)$$

$$:= \int \frac{d^D k}{i\pi^{D/2}}\frac{1}{(-k^2+m^2)^{a_1}(-(k+p_1)^2+m^2))^{a_2}(-(k+p_2)^2+m^2)^{a_3}}.$$

$$(3.238)$$

Such integrals appear for example in the process $gg \longrightarrow H$ via a top quark loop, as was discussed in Sect. 3.5.4.

We have seen above how the integration by parts reduction to master integrals works. Following the same ideas, one finds that the family of one-loop triangle integrals (with arbitrary integer powers of the propagators) can be described by a basis of three integrals, which could be canonically chosen to be $G(0, 1, 0), G(0, 1, 1), G(1, 1, 1)$. Computing them is equivalent to knowing any triangle integral within the family, thanks to the integration by parts relations.

It turns out that solving the system of differential equations becomes drastically simpler when a different basis choice is made. As we will see, this will completely clarify which special functions are required in such Feynman integrals, and the singular behavior of the functions will also become manifest.

Ideas for how to find a good choice of basis were presented in [24, 25]. Let us make an educated guess for the choice of basis, namely

$$f_1 = \frac{\varepsilon}{\Gamma(1+\varepsilon)} (m^2)^\varepsilon G(0, 2, 0), \tag{3.239}$$

$$f_2 = \frac{\varepsilon}{\Gamma(1+\varepsilon)} (m^2)^\varepsilon s \sqrt{1 - \frac{4m^2}{s}} G(0, 1, 2), \tag{3.240}$$

$$f_3 = \frac{\varepsilon^2}{\Gamma(1+\varepsilon)} (m^2)^\varepsilon s G(1, 1, 1). \tag{3.241}$$

The normalization factors were chosen such that the integrals f_i are dimensionless, and such that they start with a term ε^0 in their Laurent expansion around four dimensions. The factor $1/\Gamma(1+\varepsilon)$ was introduced for later convenience.

The differential equations then take the simple form

$$\partial_s \mathbf{f} = \varepsilon \begin{pmatrix} 0 & 0 & 0 \\ \frac{\sqrt{1-\frac{4m^2}{s}}}{4m^2-s} & \frac{1}{4m^2-s} & 0 \\ 0 & -\frac{\sqrt{1-\frac{4m^2}{s}}}{4m^2-s} & 0 \end{pmatrix} \mathbf{f}. \tag{3.242}$$

Comparing this to the differential equations obtained previously, we see the remarkable feature of Eq. (3.242) that its r.h.s. is proportional to ε. The differential equation in m^2 can be obtained in a similar way, and one may verify that

$$(s\partial_s + m^2\partial_{m^2})\mathbf{f} = 0, \tag{3.243}$$

which confirms that $\mathbf{f}(s, m^2)$ only depends on m^2/s

$$\frac{m^2}{-s} = \frac{x}{(1-x)^2}, \tag{3.244}$$

we can eliminate the square roots (by choosing e.g. $0 < x < 1$), and obtain

$$\partial_x \mathbf{f} = \varepsilon \left(\frac{a}{x} + \frac{b}{1+x} \right) \mathbf{f}, \tag{3.245}$$

with the constant matrices

$$a = \begin{pmatrix} 0 & 0 & 0 \\ 1 & 1 & 0 \\ 0 & -1 & 0 \end{pmatrix}, \qquad b = \begin{pmatrix} 0 & 0 & 0 \\ 0 & -2 & 0 \\ 0 & 0 & 0 \end{pmatrix}. \tag{3.246}$$

Equation (3.245) makes the singularity structure of f manifest. It has three singular points, $x = 0, -1, \infty$. $x = 0$ and $x = \infty$ correspond to the massless limit $m^2 \to 0$

(note that Eq. (3.244) has an $x \to 1/x$ symmetry), while $x = -1$ corresponds to the threshold limit $s = 4m^2$. Another natural point, $x = 1$, corresponds to the soft limit $s \to 0$. Note that for the integrals under consideration, the soft limit $s \to 0$ is finite and can be used as a simple boundary condition. In fact, it is easy to see that one gets

$$\mathbf{f}(x = 1, \varepsilon) = \{1, 0, 0\}. \tag{3.247}$$

Having made these general observations, we can turn to the problem of writing down the solution of the differential equations. We are interested in the perturbative solution for small ε, and therefore expand

$$\mathbf{f}(x, \varepsilon) = \sum_{k \geq 0} \varepsilon^k \mathbf{f}^{(k)}(x). \tag{3.248}$$

Plugging this expansion into Eq. (3.245), we see that it decouples order by order in ε, so that it is trivial to solve the equations. We see that the solution at order ε^k is given in terms of k-fold iterated integrals, with integration kernels either dx/x or $dx/(1 + x)$. At low orders, these functions can be explicitly written in terms of logarithms and polylogarithms, while at higher orders these are (by definition) a class of special functions called harmonic polylogarithms [26]. A useful Mathematica implementation of these functions is given in [27].

Let us spell out the first orders, taking into account the boundary condition (3.247),

$$\mathbf{f}^{(0)} = \{1, 0, 0\}, \tag{3.249}$$

$$\mathbf{f}^{(1)} = \{0, \log x, 0\}, \tag{3.250}$$

$$\mathbf{f}^{(2)} = \left\{0, \frac{1}{2} \log^2 x - 2 \log x \log(1 + x) - 2\mathrm{Li}_2(-x), -\frac{1}{2} \log^2 x\right\}. \tag{3.251}$$

In particular, taking into account the definition of f, we find agreement between $f_2^{(1)}$ and Eq. (3.233), and between $f_3^{(2)}$ and (3.235), respectively. In general, it is better to use the notation of harmonic polylogarithms, which by definition are the class of functions solving the differential equations (3.245). They have nice mathematical properties that make them easy to work with. In this way, it is straightforward to expand the solution to any desired order in ε.

Finally, let us comment on the different relevant regions for the variable x. The solution we have written down above is valid in the Euclidean region $0 < x < 1$, where the functions are real. From this region, we can analytically continue to any value of x, taking into account the $i0$ prescription of the Feynman integrals.

3.9 References and Further Reading

In this chapter we discussed a broad range of topics devoted to scattering amplitudes at the one-loop level and beyond. The topics reviewed in Sects. 3.1 to 3.3 on Feynman integrals, Wick rotation and dimensional regularization are discussed in more detail in standard textbooks on quantum field theory, including the classification of infrared divergences and the concept of dimensional regularization [28].

The possibility of reducing one-loop Feynman integrals to the basis of scalar integrals was realized first by Passarino-Veltman [3]. In Sect. 3.4 we have reviewed a rather recent and efficient reduction technique at the level of the integrand [29, 30] making extensive use of the van Neerven-Vermaseren basis [31]. A pedagogical and detailed review deepening the discussed material is [9]. The idea to construct one-loop amplitudes from their unitarity cuts, the subject of Sect. 3.5, was put forward in [11, 32]. The method of generalized unitarity employing maximal cuts to find the basis integral coefficients via tree amplitudes at complex momenta was introduced in [33] for $\mathcal{N} = 4$ super Yang-Mills theory. See [34] for an earlier application of generalized cuts. Many refinements of generalized unitarity have been developed since then, in particular towards finding the rational piece of an amplitude [35–37]. We wish to point the reader to the following reviews on generalized unitarity for further reading [9, 38–41] and also to the recent review [42].

The generalized unitarity method is highly suited for a numerical implementation in which one-loop amplitudes are calculated from trees for given numerical values of the momenta and helicities. There exist computer programs for an automated computation of one-loop amplitudes in massless QCD such as [43–49], some of which are publicly available.

In our discussion on techniques for the evaluation of loop integrals in Sects. 3.6– 3.8 we are following the conventions of Refs. [14, 15]. More examples and advanced topics are covered in this book. A further very useful reference with emphasis on Euclidean space Feynman integrals and applications to critical phenomena is [50].

The original reference for the integration-by-parts identities is [51]. The differential equation method was introduced in Refs. [52, 53], and applied to a large class of problems by [54]. For reviews, see e.g. [14, 15, 55]. There exist various useful computer algebra implementations, see e.g. [56–58]. Recently, a proposal for choosing master integrals that lead to a simpler form of the differential equations was put forward in Ref. [24]. We gave an example of this approach in Sect. 3.8.3.

In these notes we have focused mainly on techniques that are very general and therefore have a broad applicability. There are a number of other techniques for computing special classes of loop integrals that we have not covered due to lack of space. For example, some useful techniques for computing loop integrals in position space are not covered here, see [59, 60]. Another interesting topic we have not covered are dimensional recurrence relations, see e.g. [61–63].

The methods we have presented are based on analytically integrating Feynman integrals. There are also approaches to directly evaluate Feynman integrals numerically, e.g. using sector decomposition techniques, see e.g. [64–66] and references therein.

Finally, we saw that at the one-loop order we could express the analytic results in terms of logarithms and dilogarithms. Beyond the one-loop order, larger classes of special functions are needed to describe Feynman integrals analytically. One class that is frequently encountered is that of multiple polylogarithms or iterated integrals, and these functions have rich mathematical properties, see e.g. [67–69]. Useful lecture notes are available [70, 71].

References

1. G.P. Korchemsky, A.V. Radyushkin, Infrared factorization, Wilson lines and the heavy quark limit. Phys. Lett. B **279**, 359–366 (1992). arXiv:hep-ph/9203222
2. Z. Bern, A. De Freitas, L.J. Dixon, H.L. Wong, Supersymmetric regularization, two loop QCD amplitudes and coupling shifts. Phys. Rev. D **66**, 085002 (2002). arXiv:hep-ph/0202271
3. G. Passarino, M.J.G. Veltman, One loop corrections for e+ e− annihilation into mu+ mu− in the Weinberg model. Nucl. Phys. B **160**, 151 (1979)
4. R.K. Ellis, G. Zanderighi, Scalar one-loop integrals for QCD. J. High Energy Phys. **0802**, 002 (2008). arXiv:0712.1851
5. G.J. van Oldenborgh, FF: a package to evaluate one loop Feynman diagrams. Comput. Phys. Commun. **66**, 1 (1991)
6. T. Hahn, M. Perez-Victoria, Automatized one loop calculations in four-dimensions and D-dimensions. Comput. Phys. Commun. **118**, 153 (1999). arXiv:hep-ph/9807565
7. A. van Hameren, OneLOop: for the evaluation of one-loop scalar functions. Comput. Phys. Commun. **182**, 2427 (2011). arXiv:1007.4716
8. G. Cullen, J.P. Guillet, G. Heinrich, T. Kleinschmidt, E. Pilon, et al., Golem95C: a library for one-loop integrals with complex masses. Comput. Phys. Commun. **182**, 2276 (2011). arXiv:1101.5595
9. R.K. Ellis, Z. Kunszt, K. Melnikov, G. Zanderighi, One-loop calculations in quantum field theory: from Feynman diagrams to unitarity cuts. Phys. Rep. **518**, 141–250 (2012). arXiv: 1105.4319
10. R.E. Cutkosky, Singularities and discontinuities of Feynman amplitudes. J. Math. Phys. **1**, 429 (1960)
11. Z. Bern, L.J. Dixon, D.C. Dunbar, D.A. Kosower, Fusing gauge theory tree amplitudes into loop amplitudes. Nucl. Phys. B **435**, 59 (1995). arXiv:hep-ph/9409265
12. Z. Bern, L.J. Dixon, D.C. Dunbar, D.A. Kosower, One loop self-dual and $\mathcal{N} = 4$ super Yang-Mills. Phys. Lett. B **394**, 105 (1997). arXiv:hep-th/9611127
13. R.K. Ellis, W.T. Giele, Z. Kunszt, K. Melnikov, Masses, fermions and generalized D-dimensional unitarity. Nucl. Phys. B **822**, 270 (2009). arXiv:0806.3467
14. V.A. Smirnov, Evaluating Feynman integrals. Tracts Mod. Phys. **211**, 1 (2004)
15. V.A. Smirnov, Analytic tools for Feynman integrals. Tracts Mod. Phys. **250**, 1 (2012)
16. J.B. Tausk, Nonplanar massless two loop Feynman diagrams with four on-shell legs. Phys. Lett. B **469**, 225 (1999). arXiv:hep-ph/9909506
17. V.A. Smirnov, Analytical result for dimensionally regularized massless on shell double box. Phys. Lett. B **460**, 397 (1999). arXiv:hep-ph/9905323
18. M. Czakon, Automatized analytic continuation of Mellin-Barnes integrals. Comput. Phys. Commun. **175**, 559 (2006). arXiv:hep-ph/0511200
19. A.V. Smirnov, V.A. Smirnov, On the resolution of singularities of multiple Mellin-Barnes integrals. Eur. Phys. J. C **62**, 445 (2009). arXiv:0901.0386
20. J.A.M. Vermaseren, Harmonic sums, Mellin transforms and integrals. Int. J. Mod. Phys. A **14**, 2037 (1999). arXiv:hep-ph/9806280
21. S. Moch, P. Uwer, XSummer: transcendental functions and symbolic summation in form. Comput. Phys. Commun. **174**, 759 (2006). arXiv:math-ph/0508008
22. J. Ablinger, J. Blumlein, C. Schneider, Analytic and algorithmic aspects of generalized harmonic sums and polylogarithms. J. Math. Phys. **54**, 082301 (2013). arXiv:1302.0378
23. J. Gluza, K. Kajda, T. Riemann, AMBRE: a mathematica package for the construction of Mellin-Barnes representations for Feynman integrals. Comput. Phys. Commun. **177**, 879 (2007). arXiv:0704.2423

24. J.M. Henn, Multiloop integrals in dimensional regularization made simple. Phys. Rev. Lett. **110**, 251601 (2013). arXiv:1304.1806
25. J.M. Henn, A.V. Smirnov, V.A. Smirnov, Analytic results for planar three-loop four-point integrals from a Knizhnik-Zamolodchikov equation. J. High Energy Phys. **1307**, 128 (2013). arXiv:1306.2799
26. E. Remiddi, J.A.M. Vermaseren, Harmonic polylogarithms. Int. J. Mod. Phys. A **15**, 725 (2000). arXiv:hep-ph/9905237
27. D. Maitre, HPL, a mathematica implementation of the harmonic polylogarithms. Comput. Phys. Commun. **174**, 222 (2006). arXiv:hep-ph/0507152
28. G. 't Hooft, M.J.G. Veltman, Regularization and renormalization of gauge fields. Nucl. Phys. B **44**, 189 (1972)
29. G. Ossola, C.G. Papadopoulos, R. Pittau, Reducing full one-loop amplitudes to scalar integrals at the integrand level. Nucl. Phys. B **763**, 147 (2007). arXiv:hep-ph/0609007
30. W.T. Giele, Z. Kunszt, K. Melnikov, Full one-loop amplitudes from tree amplitudes. J. High Energy Phys. **0804**, 049 (2008). arXiv:0801.2237
31. W.L. van Neerven, J.A.M. Vermaseren, Large loop integrals. Phys. Lett. B **137**, 241 (1984)
32. Z. Bern, L.J. Dixon, D.C. Dunbar, D.A. Kosower, One loop n point gauge theory amplitudes, unitarity and collinear limits. Nucl. Phys. B **425**, 217 (1994). arXiv:hep-ph/9403226
33. R. Britto, F. Cachazo, B. Feng, Generalized unitarity and one-loop amplitudes in $\mathcal{N} = 4$ super-Yang-Mills. Nucl. Phys. B **725**, 275 (2005). arXiv:hep-th/0412103
34. Z. Bern, L.J. Dixon, D.A. Kosower, One-loop amplitudes for $e^+ e^-$ to four partons. Nucl. Phys. B **513**, 3 (1998). arXiv:hep-ph/9708239
35. C. Anastasiou, R. Britto, B. Feng, Z. Kunszt, P. Mastrolia, D-dimensional unitarity cut method. Phys. Lett. B **645**, 213 (2007). arXiv:hep-ph/0609191
36. R. Britto, B. Feng, Integral coefficients for one-loop amplitudes. J. High Energy Phys. **0802**, 095 (2008). arXiv:0711.4284
37. S.D. Badger, Direct extraction of one loop rational terms. J. High Energy Phys. **0901**, 049 (2009). arXiv:0806.4600
38. Z. Bern, L.J. Dixon, D.A. Kosower, On-shell methods in perturbative QCD. Ann. Phys. **322**, 1587–1634 (2007). arXiv:0704.2798
39. R. Britto, Loop amplitudes in gauge theories: modern analytic approaches. J. Phys. A **44**, 454006 (2011). arXiv:1012.4493
40. Z. Bern, Y.-t. Huang, Basics of generalized unitarity. J. Phys. A **44**, 454003 (2011). arXiv:1103.1869
41. L.J. Dixon, A brief introduction to modern amplitude methods (2013). arXiv:1310.5353
42. H. Elvang, Y.-t. Huang, Scattering amplitudes (2013). arXiv:1308.1697
43. C.F. Berger, et al., An automated implementation of on-shell methods for one-loop amplitudes. Phys. Rev. D **78**, 036003 (2008). arXiv:0803.4180
44. W.T. Giele, G. Zanderighi, On the numerical evaluation of one-loop amplitudes: the gluonic case. J. High Energy Phys. **0806**, 038 (2008). arXiv:0805.2152
45. P. Mastrolia, G. Ossola, T. Reiter, F. Tramontano, Scattering AMplitudes from unitarity-based reduction algorithm at the integrand-level. J. High Energy Phys. **1008**, 080 (2010). arXiv:1006.0710
46. S. Badger, B. Biedermann, P. Uwer, NGluon: a package to calculate one-loop multi-gluon amplitudes. Comput. Phys. Commun. **182**, 1674 (2011). arXiv:1011.2900
47. G. Bevilacqua, M. Czakon, M.V. Garzelli, A. van Hameren, A. Kardos, et al., HELAC-NLO. Comput. Phys. Commun. **184**, 986 (2013). arXiv:1110.1499
48. G. Cullen, N. Greiner, G. Heinrich, G. Luisoni, P. Mastrolia, et al., Automated one-loop calculations with GoSam. Eur. Phys. J. C **72**, 1889 (2012). arXiv:1111.2034
49. S. Badger, B. Biedermann, P. Uwer, V. Yundin, Numerical evaluation of virtual corrections to multi-jet production in massless QCD. Comput. Phys. Commun. **184**, 1981–1998 (2013). arXiv:1209.0100
50. H. Kleinert, V. Schulte-Frohlinde, *Critical Properties of phi**4-Theories* (Word Scientific, Singapore, 2001)

51. K.G. Chetyrkin, F.V. Tkachov, Integration by parts: the algorithm to calculate beta functions in 4 loops. Nucl. Phys. B **192**, 159 (1981)
52. A.V. Kotikov, Differential equations method: new technique for massive Feynman diagrams calculation. Phys. Lett. B **254**, 158 (1991)
53. A.V. Kotikov, Differential equation method: the calculation of N point Feynman diagrams. Phys. Lett. B **267**, 123 (1991)
54. T. Gehrmann, E. Remiddi, Differential equations for two loop four point functions. Nucl. Phys. B **580**, 485 (2000). arXiv:hep-ph/9912329
55. M. Argeri, P. Mastrolia, Feynman diagrams and differential equations. Int. J. Mod. Phys. A **22**, 4375 (2007). arXiv:0707.4037
56. C. Anastasiou, A. Lazopoulos, Automatic integral reduction for higher order perturbative calculations. J. High Energy Phys. **0407**, 046 (2004). arXiv:hep-ph/0404258
57. A.V. Smirnov, Algorithm FIRE—Feynman integral REduction. J. High Energy Phys. **0810**, 107 (2008). arXiv:0807.3243
58. C. Studerus, Reduze-Feynman integral reduction in C++. Comput. Phys. Commun. **181**, 1293 (2010). arXiv:0912.2546
59. K.G. Chetyrkin, A.L. Kataev, F.V. Tkachov, New approach to evaluation of multiloop Feynman integrals: the Gegenbauer polynomial x space technique. Nucl. Phys. B **174**, 345 (1980)
60. D.I. Kazakov, The method of uniqueness, a new powerful technique for multiloop calculations. Phys. Lett. B **133**, 406 (1983)
61. O.V. Tarasov, Connection between Feynman integrals having different values of the space-time dimension. Phys. Rev. D **54**, 6479 (1996). arXiv:hep-th/9606018
62. R.N. Lee, Space-time dimensionality D as complex variable: calculating loop integrals using dimensional recurrence relation and analytical properties with respect to D. Nucl. Phys. B **830**, 474 (2010). arXiv:0911.0252
63. R.N. Lee, A.V. Smirnov, V.A. Smirnov, Dimensional recurrence relations: an easy way to evaluate higher orders of expansion in ε. Nucl. Phys., Proc. Suppl. **205–206**, 308 (2010). arXiv:1005.0362
64. C. Bogner, S. Weinzierl, Resolution of singularities for multi-loop integrals. Comput. Phys. Commun. **178**, 596 (2008). arXiv:0709.4092
65. A.V. Smirnov, M.N. Tentyukov, Feynman integral evaluation by a sector decomposition approach (FIESTA). Comput. Phys. Commun. **180**, 735 (2009). arXiv:0807.4129
66. J. Carter, G. Heinrich, SecDec: a general program for sector decomposition. Comput. Phys. Commun. **182**, 1566 (2011). arXiv:1011.5493
67. L. Lewin, Structural properties of polylogarithms
68. K.-T. Chen, Iterated path integrals. Bull. Am. Math. Soc. **83**(5), 831 (1997)
69. A.B. Goncharov, Multiple polylogarithms, cyclotomy and modular complexes. Math. Res. Lett. **5**, 497 (1998). arXiv:1105.2076
70. F. Brown, Iterated integrals in quantum field theory. http://www.math.jussieu.fr/brown/
71. J. Zhao, Multiple polylogarithms (2013). http://www.maths.dur.ac.uk/events/Meetings/LMS/2013

Chapter 4
Advanced Topics

In this chapter we discuss advanced topics closely related to current research. The study of scattering amplitudes is a quickly evolving field, and it is therefore impossible to cover all developments or give a complete up-to-date review here. Having said that, we have therefore made a selection that seemed suitable for these lecture notes. Some of the sections can be used independently for additional lectures, or can provide a basis for student presentations.

In Sect. 4.1 we discuss recent ideas generalizing the use of on-shell recursion relations from tree-level amplitudes to loop-level integrands. We give a pedagogical example at one loop in $\mathcal{N} = 4$ super Yang-Mills. The remaining sections focus on topics in $\mathcal{N} = 4$ super Yang-Mills. We give an introduction to a surprising duality between Wilson loops and scattering amplitudes in Sect. 4.3. Appendix A provides the necessary background regarding renormalization properties of Wilson loops. The consequences of conformal symmetry for the Wilson loops is discussed in Sect. 4.4. For the amplitudes, it implies the existence of a dual conformal symmetry that is part of a larger Yangian symmetry, discussed in Sect. 4.5.

4.1 Recursion Relations for Loop Integrands

In previous chapters we discussed in detail the properties of tree-level scattering amplitudes and showed that they satisfy powerful recursion relations. The latter allow to construct arbitrary n-point amplitudes starting only from basic on-shell three-particle amplitudes. The key idea was to consider the amplitudes as analytic functions of the external momenta. Introducing certain complex shifts of the latter, one can define a complex deformation $\mathscr{A}(z)$ of the scattering amplitudes. The poles at special values of z correspond to internal propagators going on-shell. Near such kinematical configurations, the amplitudes have simple factorization properties. In this way, the calculation of an arbitrary tree-level amplitude is recursively related to the elementary on-shell three-point vertices.

J.M. Henn, J.C. Plefka, *Scattering Amplitudes in Gauge Theories*,
Lecture Notes in Physics 883, DOI 10.1007/978-3-642-54022-6_4,
© Springer-Verlag Berlin Heidelberg 2014

Let us discuss the extension of these concepts to loop level. An l-loop scattering amplitude of n external legs can in general be written as

$$\mathscr{A}(p_1, \ldots, p_n) = \int d^D k_1 \cdots \int d^D k_l \, \mathscr{I}(p_1, \ldots, p_n; k_1, \ldots, k_l), \qquad (4.1)$$

where $\mathscr{I}$ is a rational function, the *loop integrand*. Computing $\mathscr{A}$ can be viewed in terms of two steps: first, to obtain the integrand $\mathscr{I}$ in a suitable form, and second, to perform the integrations. The second step, usually the more complicated one, was discussed in Chap. 3. As for the first step, in principle $\mathscr{I}$ is determined by the Feynman rules of the theory. In many situations, this can be the approach of choice, especially since many automated computer codes exist [1–3]. However, the computation can be cumbersome, for the same reasons that we discussed at tree-level, and it is worthwhile having a conceptually simpler alternative at hand. This also allows to understand the physical properties of integrands and amplitudes in a transparent way.

The integrated amplitude $\mathscr{A}$ is typically a complicated function involving logarithms, polylogarithms, or other special functions having branch cuts. The latter severely complicate the straightforward application of the idea of recursion relations. It is possible to use recursion relations at the level of the integrated functions, but we will not discuss them here. We refer the interested reader to Ref. [4]. On the other hand, setting ourselves the more modest goal of obtaining the loop integrand $\mathscr{I}$ is easier. Indeed, the latter has properties very similar to that of tree-level amplitudes discussed earlier. It is a rational function that can be defined using Feynman rules. The difference to tree-level is that the topology of the Feynman graphs is now more general, because graphs forming loops are also allowed.

The key idea is therefore to use recursion relations in order to determine $\mathscr{I}$, bypassing Feynman diagrams. Although tree-level amplitudes and loop integrands are very similar, there are a few subtleties for loop integrands that need to be kept in mind. They are all related to the proper definition of what we mean by *'the' loop integrand*, i.e. the uniqueness of its definition. Let us discuss them in turn.

- One issue is that the choice of loop momenta in Eq. (4.1) is arbitrary. For instance, one could define a different loop integrand by performing a shift $k_i^\mu \longrightarrow k_i^\mu + \Delta_i^\mu$. For planar amplitudes, this ambiguity can be removed by introducing dual coordinates, which corresponds to a specific and unique choice of loop momenta.
- A related point is that two integrands that differ only by total derivatives lead to the same integrated amplitude. One way of avoiding this ambiguity is to use Feynman rules as a primary definition of the loop integrand.
- In theories with UV divergences, one has to take into account wavefunction renormalization when defining the S-matrix elements. Recall that contributions from bubble-type insertions on external legs are 'amputated' in the LSZ definition of the loop level S-matrix, as was reviewed in Sect. 1.4. The corresponding Feynman diagrams need to be treated with care, since such contributions are singular when going on-shell. See Ref. [5] for further discussion of this point.

All the above subtleties are of course important and of practical relevance. However, in order to show the main idea of the recursion relations, let us ignore them for now, by proceeding in a simple setting where they do not arise. Specifically, let us consider planar scattering amplitudes in a theory without UV divergences, such as $\mathcal{N} = 4$ super Yang-Mills. For the extension to more general theories, the above points will have to be taken into account. Having said this, let us now discuss the recursion relations.

Just as at tree-level, we can study BCFW shifts of the external momenta that introduce a new complex parameter z, such that $\mathscr{I} \longrightarrow \mathscr{I}(z)$. Specifically, we take

$$\lambda_1 \longrightarrow \lambda_1 - z\lambda_n, \tag{4.2}$$

$$\tilde{\lambda}_n \longrightarrow \tilde{\lambda}_n + z\tilde{\lambda}_1, \tag{4.3}$$

$$\eta_n \longrightarrow \eta_n + z\eta_1, \tag{4.4}$$

as done before in Eqs. (2.140) and (2.143). Note that it is also possible to consider shifts involving the loop momenta, but we will not discuss them here.

It is clear from the structure of the underlying Feynman graphs that similar factorization properties as for tree-level amplitudes hold. There are two types of poles in the complex z plane that can occur, depending on whether the on-shell propagator is part of a loop or not. In the latter case, we see that the residue of the pole factorizes into lower-point integrands. So this situation is completely analogous to the tree-level case. A new feature occurs in the former case. Taking the propagator on-shell can be viewed as 'cutting' the corresponding loop open, and one obtains an amplitude with one loop less, but two additional external legs. Moreover, because of momentum conservation, the lower-loop amplitude is in a special kinematical configuration, the forward limit, i.e. the cut lines have momenta k^μ and $-k^\mu$, respectively. This limit is well-defined and finite for supersymmetric field theories [5]. The two types of contributions are displayed on the r.h.s. of the equation in Fig. 4.1.

In summary, we see that, at least in supersymmetric theories, one can apply the recursion relations to compute loop integrands. All terms on the r.h.s. of the recursion relations for an l-loop N^kMHV n-point integrand correspond to other integrands, with either l, k, or n is lowered. This means that, just as at tree-level, the recursion can be systematically solved, starting from the on-shell tree-level three-point vertices. This recursion is displayed pictorially in Fig. 4.1. For example, in this approach, the one-loop four-point MHV integrand is obtained via the forward limit of a six-point tree-level NMHV amplitude. This will be discussed at the end of this section.

Let us comment on some of the practical aspects of this approach. Of course, the recursion relations are most easily solved in the massless four-dimensional case, and the inclusion of masses or the treatment in D dimensions is technically more challenging. However, even though the integrand by itself is finite in $D = 4$ dimensions, in the context of dimensional regularization it is usually necessary to know the latter in D dimensions, in order for the subsequent integration to be well-defined. The importance of this can be seen for example in the so-called 'rational terms' in one-loop

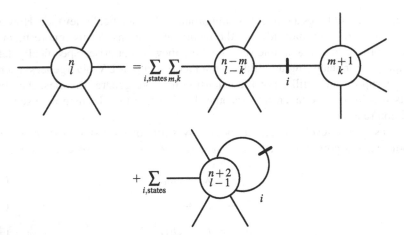

Fig. 4.1 Pictorial representation of the on-shell recursion relations for an l-loop n-point loop integrand. The first term on the r.h.s. arises from a poles of propagators that are not part of a loop, as at tree-level. In the second term on the r.h.s. the pole arises from a propagator within a loop. This term is related to the forward limit of a lower-loop amplitude with two additional legs

amplitudes, discussed in Sect. 3.5.3, which naively have a zero integrand, but do lead to a non-zero answer when the regularization prescription is followed consistently. More generally, technically the mismatch between the four-dimensional external states in the spinor-helicity formalism and the D-dimensional loop momenta introduces a dependence on the dimension. One may view the integrand as an expansion, where the four-dimensional term is the first one. Of course, this issue also arises in the unitarity approach, where it is also vastly simpler to work with four-dimensional cuts. See [6] for a discussion of unitarity methods in D dimensions.

The recursion relations typically give the integrand in non-local form, i.e. in a form that may contain spurious singularities. At present it is not yet known how to efficiently integrate such forms directly, such that one usually rewrites these expression in terms of more standard local integrands.

In summary, recursion relations are a conceptionally elegant way of describing and computing loop integrands. They constitute a natural generalization of the BCFW recursion relations at tree-level. The method is also closely related to generalized unitarity cuts, in the sense that all information is contained in on-shell objects. The difference is that the recursive approach is constructive, such that it does not rely on knowing the general form of the answer or a loop integral basis a priori.

So far, the bulk of applications of the recursion relations has been to the four-dimensional part of loop integrands in planar $\mathcal{N} = 4$ super Yang-Mills [7]. Important open questions include the application of this method for theories with no supersymmetry, or in non-planar cases.

Example: Four-Point One-Loop Integrand in $\mathcal{N} = 4$ Super-Yang-Mills   Here we give a simple example of the application of the integrand recursion relations in $\mathcal{N} = 4$ super Yang-Mills, for the four-point one-loop integrand. We also discussed

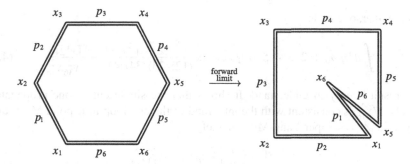

Fig. 4.2 Kinematic configuration of the six-particle amplitude approaching the forward limit $p_5 \approx p_6$. In the limiting configuration dual points x_1 and x_5 are identified and x_6 plays the role of the integration variable in the corresponding one-loop integrand

this case using generalized unitarity methods in Sect. 3.5. Let us analyze the two types of terms on the r.h.s. of the equation in Fig. 4.1. For four external legs the first term vanishes, since the loop-level three-point integrand vanishes, even for complex momenta. The second term involves the forward limit of the six-particle NMHV amplitude, with the intermediate states that are exchanged across the cut being summed over,

$$\mathscr{I}_4 \propto \frac{1}{x_{16}^2} \int d^4\eta \mathscr{A}_6(\hat{1}, 2, 3, \hat{4}, 5, 6)|_{\text{forward}}. \tag{4.5}$$

Here the hats denote the BCFW shift, and 'forward' indicates that the amplitude is evaluated in the forward limit $p_6 = -p_5$, see Fig. 4.2. In spinor helicity language this can be realized by setting $\lambda_6 = \lambda_5$, $\tilde{\lambda}_6 = -\tilde{\lambda}_5$, and $\eta_6 = -\eta_5 \equiv \eta$. The common η variable is being integrated over in Eq. (4.5) in order to reproduce all possible particle exchanges across the cut (gluons, fermions, scalars). As we will see below, the forward limit is somewhat subtle and has to be taken with care.

Let us evaluate the amplitude $\mathscr{A}_6$ in the forward limit. Our starting point is the six-particle NMHV amplitude, which can be written as, cf. Eq. (2.176),

$$\mathscr{A}_6^{\text{NMHV}} \propto \delta^{(8)}\left(\sum_{i=1}^{6} \lambda_i \eta_i\right) \frac{1}{\langle 12 \rangle \cdots \langle 56 \rangle \langle 61 \rangle}[R_{6;24} + R_{6;25} + R_{6;35}]. \tag{4.6}$$

It is obvious from the presence of $\langle 56 \rangle \to 0$ in the denominator of Eq. (4.6) that the forward limit is subtle, and needs to be parametrized carefully. As we will see, it is singular for the component amplitudes, and is rendered finite by the sum over intermediate state. Performing the integral $\int d^4\eta$ over the common η integral yields a factor of $\langle 56 \rangle^4$. Therefore we can only obtain a non-zero answer in the forward limit if there are corresponding singular factors in the expression for $R_{n;st}$. Analyzing them one can see that the terms corresponding to $R_{6;24}$ and $R_{6;25}$ yield a vanishing contribution, and that $R_{6;35}$ leads to a finite contribution. Putting all factors together, one finds

$$\int d^4\eta \mathscr{A}_6(1,2,3,4,5,6)|_{\text{forward}} \propto \frac{\delta^{(8)}(\sum_{i=1}^4 \lambda_i \eta_i)}{\langle 12\rangle\langle 23\rangle\langle 34\rangle\langle 41\rangle} \frac{x_{13}^2 x_{24}^2}{x_{26}^2 x_{36}^2 x_{46}^2}. \tag{4.7}$$

This result is easy to understand. It shows that the single cut of one propagator, namely $1/x_{16}^2$ is consistent with the integrand of the one-loop four-point MHV amplitude in $\mathscr{N} = 4$ super Yang-Mills, namely

$$\mathscr{I}_4 \propto \frac{\delta^{(8)}(\sum_{i=1}^4 \lambda_i \eta_i)}{\langle 12\rangle\langle 23\rangle\langle 34\rangle\langle 41\rangle} \frac{x_{13}^2 x_{24}^2}{x_{16}^2 x_{26}^2 x_{36}^2 x_{46}^2}, \tag{4.8}$$

where x_6 represents the dual integration point. This is in the spirit of the (generalized) unitarity approach of Sect. 3.5. However, in the context of the recursion relations, this equation gives us more than a consistency check. It gives a derivation of the four-dimensional integrand, by evaluating Eq. (4.5). In the present case, the shift leaves the tree-level amplitude $\mathscr{A}_6$ in the formula above invariant, and we immediately arrive at the integrand given in Eq. (4.8). As mentioned before, the above derivation is valid for the $D = 4$ dimensional part of the integrand, and additional arguments have to be made in order to constrain potential deviations from four dimensions. In the present case, one can exclude additional terms e.g. by evaluating two-particle cuts in D-dimensions.

This completes our example. Let us conclude this section with a number of comments. In more complicated cases, the result of the BCFW recursion gives, just as at tree-level, results in a non-local form that contains spurious poles. One can rewrite the expressions so obtained in local fashion that is more suitable to traditional integration methods. A final comment is that the recursion relations in $\mathscr{N} = 4$ super Yang-Mills are even more conveniently written in (momentum) twistor variables [8, 9]. These variables are also natural in the context the symmetry properties of scattering amplitudes, see Sect. 4.5.

Exercise 4.1 (Forward Limit of the NMHV Tree-Level Amplitude in $\mathscr{N} = 4$ Super Yang-Mills)
    Verify the calculation of the forward limit of the NMHV tree-level amplitude given in the main text. In particular, show that the contribution of $R_{6;24}$ and $R_{6;35}$ to Eq. (4.7) vanishes, and that the contribution of $R_{6;25}$ correctly reproduces its r.h.s.

4.2 Scattering Amplitudes in $\mathscr{N} = 4$ Super Yang-Mills

In the previous chapter we discussed methods applicable for scattering amplitudes in generic Yang-Mills theories. We have already mentioned that in the supersymmetric case simplifications arise, for example due to better UV properties. The distinguished case of maximally supersymmetric Yang-Mills theory, $\mathscr{N} = 4$ super Yang-Mills, is in fact UV finite. Being a massless quantum field theory with zero beta function, it has a quantum conformal symmetry. In addition, it has many special properties, including a conjectured duality to string theory, the AdS/CFT correspondence.

Why should one study scattering amplitudes in a conformal field theory that is not realized in Nature? One can argue that studying Yang-Mills theories with varying amount of supersymmetry allows one to understand better the structure of gauge theories. Perturbative calculations, e.g. for scattering amplitudes, require similar methods, and often insights gleaned from the (easier) supersymmetric case lead to applications in QCD as well. An example are the analytic formulas for massless tree-level scattering amplitudes in QCD, which are being used in numerical phenomenological predictions for cross-sections. They were obtained by first solving the supersymmetric case, which in turn was facilitated by a hidden Yangian symmetry that will be discussed in this chapter, and then studying the necessary modifications for the non-supersymmetric case.

The strategy of gleaning insights from the supersymmetric theories, which enabled us to solve the tree-level problem completely, is also very useful at loop level. For example, it allows one to untangle effects due to ultraviolet and infrared divergences. Although both are in principle understood, they introduce significant practical difficulties.

Work in those directions has already lead to new techniques which have applications beyond $\mathcal{N} = 4$ SYM. In this chapter, we will present some of the surprising new results found for scattering amplitudes in $\mathcal{N} = 4$ SYM. We will begin by reviewing a duality between Wilson loops and scattering amplitudes, which will help to motivate a hidden symmetry for scattering amplitudes. Choosing appropriate variables, the latter symmetry can be made manifest for scattering amplitudes, and it explains the simplicity of tree-level scattering amplitudes, including the ones in pure Yang-Mills theory.

4.3 Wilson Loop/Scattering Amplitude Duality

We have already mentioned that $\mathcal{N} = 4$ SYM is a field theory with many special properties. One of the surprising properties of this theory that emerged over the last years is a duality between scattering amplitudes and certain polygonal Wilson loops. There exist several reviews on this topic [10, 11]. Here we will illustrate the main points of the duality in the MHV case. As an example we will discuss the four-point case in detail. This section assumes some familiarity with Wilson loops and their renormalization properties. Some of these basic properties of Wilson loops are reviewed in Appendix A.

Duality Between MHV Amplitudes and Bosonic Wilson Loops Recall that we can write a general n-point MHV amplitude in the following factorized form,

$$\mathscr{A}_n^{\mathrm{MHV}} = \mathscr{A}_n^{\mathrm{MHV,tree}} \times M_n, \tag{4.9}$$

Fig. 4.3 Relation between on-shell momenta of the scattering amplitudes and dual coordinates (or region momenta)

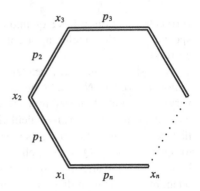

where the loop corrections M_n only depend on the momenta p_i, but not on the helicities of the particles. The object dual to these amplitudes are Wilson loops defined on n-sided polygonal contours. The latter live in a dual coordinate space x_i. It is related to the momenta via (cf. Sect. 3.6.1)

$$x_{i+1} - x_i = p_i \quad (\text{mod } n). \tag{4.10}$$

Because of momentum conservation, the p_i define a closed contour in the auxiliary coordinate space of the x_i. See Fig. 4.3. Note in particular that the x_i are not the Fourier transforms of the p_i. The on-shell conditions $p_i^2 = 0$ imply that consecutive points on the contour are light-like (i.e. null-separated), $x_{i,i+1}^2 = 0$.

The duality between MHV amplitudes and Wilson loops, illustrated in Fig. 4.4, can then be written as follows,

$$\log M_n + \mathcal{O}(1/N) \longleftrightarrow \log\langle W_n \rangle. \tag{4.11}$$

Here the arrows mean that the suitably defined finite part of the objects on the l.h.s. and r.h.s. are equal. Perhaps a more satisfactory way to state the duality is in terms of regularization-independent quantities. The key point is that both the IR divergences of scattering amplitudes, as well as the UV divergences of Wilson loops have a universal form. What we are comparing is a regulator-independent 'hard function', independent of the regularization scheme. Such a function can be obtained, for example, by taking ratios of appropriately chosen Wilson loop correlators/amplitudes.

In order to understand such ratios better, let us first review the structure of UV divergences of Wilson loops with cusps (see Appendix A for more details). For a Wilson loop C_n formed by n light-like segments connecting points x_i, the ultraviolet divergences take a factorized form,

$$\log\langle W[C_n]\rangle = \sum_{i=1}^{n} D\big(x_{i-1,i+1}^2\big) + F_n\big(x_{ij}^2\big), \tag{4.12}$$

with the finite part F_n being a function of dimensionless ratios of the (nonzero) Lorentz invariants x_{ij}^2. This formula tells us that the cusp divergences are local,

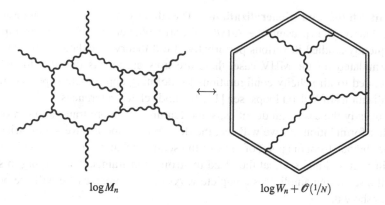

$$\log M_n \qquad\qquad \log W_n + \mathcal{O}(1/N)$$

Fig. 4.4 Duality between scattering amplitudes and Wilson loops. Up to non-planar corrections, suitably defined finite parts of the (logarithm of) scattering amplitudes and Wilson loops are equal to each other. The on-shell momenta of the scattering amplitudes are related to the coordinate space variables x_i via $x_{i+1} - x_i = p_i \pmod{n}$

i.e. the divergences corresponding to the cusp at x_i^μ only depend on the (light-like separated) neighboring points x_{i-1}^μ and x_{i+1}^μ, and can be removed by a multiplicative renormalization. In planar $\mathcal{N} = 4$ super-Yang Mills, when using a supersymmetric regularization scheme, the cusp divergences are expected to have the form

$$D(x^2) = -\frac{1}{4}\sum_{\ell \geq 1} a^\ell (-x^2 \mu^2)^{\ell\varepsilon}\left(\frac{\Gamma_{\text{cusp}}^{(\ell)}}{(\ell\varepsilon)^2} + \frac{\Gamma^{(\ell)}}{\ell\varepsilon}\right), \qquad a = \frac{g^2 N}{8\pi^2}. \qquad (4.13)$$

Here $\Gamma_{\text{cusp}}^{(\ell)}$ and $\Gamma^{(\ell)}$ are coefficients of anomalous dimensions. In particular, the planar cusp anomalous dimension [12] $\Gamma = \sum_{\ell > 0} a^\ell \Gamma_{\text{cusp}}^{(\ell)}$ is conjectured to be known to all loop orders in $\mathcal{N} = 4$ super Yang-Mills [13].

From Eqs. (4.12) and (4.13) we see that the ultraviolet divergences of Wilson loops factorize into pieces that depend on one Mandelstam variable $x_{i,i+2}^2$ at a time. The form of the divergences is universal and does not depend on n. Therefore, we can define the following infrared finite quantity, for $n \geq 4$,

$$H_n^{\text{WL}} = \log\left(\frac{W_n(\{x_{ij}^2\})}{[\prod_{i=1}^n W_4(x_{i,i+2}^2, x_{i,i+2}^2)]^{1/4}}\right) + \mathcal{O}(1/N). \qquad (4.14)$$

Here $W_n(\{x_{ij}^2\})$ is a generic n-point Wilson loop, whereas W_4 is evaluated for its two arguments being equal, in order to reproduce the ultraviolet divergences of W_n in the different channels. We can make an analogous definition for the scattering amplitudes, H_n^{A}, keeping only the leading planar term in each case. Then the statement of the duality simply is

$$H_n^{\text{WL}} = H_n^{\text{A}}. \qquad (4.15)$$

Current Status and Generalizations The duality was initially suggested at strong coupling [14], using the AdS/CFT correspondence. This relationship was subsequently studied in various perturbative field theory calculations. Initially being formulated for the MHV case, there are now arguments that the duality can be extended to all helicity configurations by defining appropriate supersymmetric generalizations of Wilson loops, see [15, 16] and related references.

As usually the case with dualities, some features are more transparent in one or the other formulation. As we will see, the dual Wilson loops have symmetries that are not obvious from the point of view of the scattering amplitudes. We will discuss them in Sect. 4.4 and see that they lead to strong constraints, determining in some cases the scattering amplitudes completely. As one example, as we will see below, one can show that

$$H_4 = \frac{1}{4} \Gamma_{\mathrm{cusp}}(a) \log^2 \frac{x_{13}^2}{x_{24}^2}, \qquad (4.16)$$

to all orders in the coupling constant a. In the following example, we will see this in detail at one loop, and explain the general proof in the following section.

Example: Four-Sided Wilson Loop Consider a polygonal contour C_4 with corners $x_1^\mu, \ldots, x_4^\mu$, such that the edges $x_{12}^2 = x_{23}^2 = x_{34}^2 = x_{41}^2 = 0$. The Wilson loop vacuum expectation value for such a contour will depend on the dimensionful variables x_{13}^2 and x_{24}^2, the UV renormalization scale μ, and the dimensional regularization parameter ε, where $D = 4 - 2\varepsilon$, with $\varepsilon > 0$ for UV divergences.

We would like to compute the vacuum expectation value $\langle W[C_4] \rangle$ of the Wilson loop defined on this contour, i.e.

$$\langle W[C_4] \rangle = \frac{1}{N} \langle \mathrm{Tr} P e^{ig \int_C ds \dot{x}^\mu A_\mu(x(s))} \rangle. \qquad (4.17)$$

What will be the structure of the perturbative results? Looking at Eqs. (4.12) and (4.13), we expect double poles in the UV regulator ε at one loop. Moreover, for our case of $n = 4$ points, the finite part $F_4 = F(x_{13}^2/x_{24}^2)$ is a function of one dimensionless variable.

Let us now perform the calculation of the one-loop vacuum expectation value. Our goal will be to verify the above structure of UV divergences, and to compute the finite part. We will require the coordinate space gluon propagator at lowest order. It reads, in Feynman gauge,

$$G^{\mu\nu}(x) = \langle A^\mu(x) A^\nu(0) \rangle = -\eta^{\mu\nu} \frac{\Gamma(1-\varepsilon)}{4\pi^2} (-x^2 + i0)^{-1+\varepsilon} (\mu^2 e^{-\gamma_E})^\varepsilon, \quad (4.18)$$

where we have redefined the canonical dimensional regularization scale $\pi^2 \mu^2 \rightarrow \mu^2 e^{-\gamma_E}$ for convenience. Up to the one-loop order, one obtains

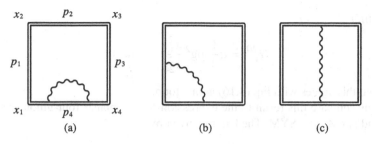

Fig. 4.5 The three types of Feynman diagrams arising in the one-loop computation of a four-sided Wilson loop

$$\langle W[C_4]\rangle = 1 + \frac{1}{2}(ig)^2 C_F \int_C dx^\mu \int_C dy^\mu\, G_{\mu\nu}(x - y) + \mathcal{O}(g^4). \qquad (4.19)$$

There are three types of diagrams to be considered, shown in Fig. 4.5. Note that the answer is gauge invariant, but individual diagrams are not. Let us discuss them in turn.

Diagram (a) is somewhat subtle due to the light-likeness of the lines. One can argue that it has to vanish because it does not depend on any scale. For a more careful argument, one could start with a contour where $x_{i,i+1}^2 = p^2 \neq 0$, in $4 - 2\varepsilon$ dimensions, and take the limit $p^2 \to 0$. The cusp diagram (b), has both soft and collinear divergences, which lead to double poles in the dimensional regularization parameter ε. Diagram (c), on the other hand, is finite, and can be computed in $D = 4$ dimensions. We leave the computation as an exercise to the reader (see exercise at the end of this section).

One arrives at [17]

$$\log\langle W[C_4]\rangle = a\left\{-\frac{1}{\varepsilon^2}\left[\left(\frac{\mu_{\mathrm{UV}}^2}{-x_{13}^2}\right)^{-\varepsilon_{\mathrm{UV}}} + \left(\frac{\mu_{\mathrm{UV}}^2}{-x_{24}^2}\right)^{-\varepsilon_{\mathrm{UV}}}\right]\right.$$

$$\left. + \frac{1}{2}\log^2\frac{x_{13}^2}{x_{24}^2} + \frac{\pi^2}{3} + \mathcal{O}(\varepsilon)\right\} + \mathcal{O}(a^2). \qquad (4.20)$$

We may verify that this is in agreement with the expected form Eqs. (4.12), (4.13), with the values

$$\Gamma_{\mathrm{cusp}}^{(1)} = 2, \qquad \Gamma^{(1)} = 0, \qquad F_4^{(1)} = \frac{1}{2}\log^2\frac{x_{13}^2}{x_{24}^2} + \frac{\pi^2}{3}. \qquad (4.21)$$

In order to compare to the four-particle scattering amplitude, let us evaluate the regulator-independent hard function H_4^{WL} defined in Eq. (4.14). For $n = 4$, it simply becomes

$$H_4^{\mathrm{WL}} = \log\left(\frac{\langle W(x_{13}^2, x_{24}^2)\rangle}{\sqrt{\langle W(x_{13}^2, x_{13}^2)\rangle\langle W(x_{24}^2, x_{24}^2)\rangle}}\right) = F_4\left(x_{13}^2/x_{24}^2\right) - F_4(1). \qquad (4.22)$$

We find

$$H_4^{\text{WL}} = a\frac{1}{2}\log^2\frac{x_{13}^2}{x_{24}^2} + \mathcal{O}(a^2).\tag{4.23}$$

Note that this agrees with Eq. (4.16) at one loop.

Let us compare this result to the expression for the planar four-gluon scattering amplitude in $\mathcal{N} = 4$ SYM. The latter is given by

$$A_4 = A_4^{\text{tree}} M_4,\tag{4.24}$$

where (cf. Sect. 3.6)

$$M_4 = 1 + a\left\{\frac{1}{\varepsilon^2}\left[\left(\frac{\mu^2}{-s}\right)^\varepsilon + \left(\frac{\mu^2}{-t}\right)^\varepsilon\right] + \frac{1}{2}\log^2\frac{s}{t} + \frac{2}{3}\pi^2 + \mathcal{O}(\varepsilon)\right\} + \mathcal{O}(a^2).\tag{4.25}$$

Here the precise form of the infrared divergent terms and the constant in the finite part are inessential, as they drop out in the definition of H_4^A. We find

$$H_4^A = a\frac{1}{2}\log^2\frac{s}{t} + \mathcal{O}(a^2)\tag{4.26}$$

in perfect agreement with the duality relation (4.15), upon the identification $x_{13}^2 = s, x_{24}^2 = t$.

Exercise 4.2 (Calculation of Four-Cusped Wilson Loop at One Loop)

Compute the Feynman diagrams shown in Fig. 4.5 in Feynman gauge, and verify the final expression given in Eq. (4.20).

Hint 1: For four general points $x_a^\mu, x_b^\mu, x_c^\mu, x_d^\mu$, one has $2x_{ab}\cdot x_{cd} = x_{ad}^2 + x_{bc}^2 - x_{ac}^2 - x_{bd}^2$.

Hint 2: This calculation is presented in detail in Sect. 4 of Ref. [17].

4.4 (Dual) Conformal Symmetry

4.4.1 Conformal Ward Identities for Cusped Wilson Loops

The polygonal light-like Wilson loops we discussed in the previous paragraphs have a conformal symmetry that can be used to constrain their vacuum expectation value. In fact, while conformal transformations change the coordinates x_i^μ, they leave light-like separations invariant. Since the Wilson loops are computed in a conformally invariant field theory, one would naively conclude that they are invariant under such transformations. However, due to UV divergences the calculations have to be carried out in the presence of a regulator, and the dimensional regulator we chose breaks this symmetry.

Nevertheless, one can still study the consequences of this broken symmetry, especially as the regulator ε tends to zero. Due to the local nature of UV divergences, the terms that contribute in this limit are under control, and a careful analysis leads to the following Ward identity [18],

$$K^\mu F_n = \frac{1}{2} \Gamma_{\text{cusp}}(a) \sum_{i=1}^{n} x_{i,i+1}^\mu \log\left(\frac{x_{i,i+2}^2}{x_{i-1,i+1}^2}\right), \tag{4.27}$$

where

$$K^\mu = \sum_{i=1}^{\mu} \left[2x_i^\mu x_i^\nu \frac{\partial}{\partial x_{i\nu}} - x_i^2 \frac{\partial}{\partial x_{i\mu}} \right], \tag{4.28}$$

is the generator of infinitesimal special conformal transformations, cf. Exercise 2.7. Note that the dependence of the r.h.s. of Eq. (4.27) on the coupling constant enters only through the cusp anomalous dimension $\Gamma_{\text{cusp}}(a)$. The reason is that the anomaly term is due to UV divergences.

What are the consequences of Eq. (4.27)? It is solved by a particular solution, plus solutions to the homogeneous equation. The latter can be any function of conformally invariant cross-ratios of the type $x_{ij}^2 x_{mn}^2 / (x_{in}^2 x_{jm}^2)$. However, due to the light-likeness of the contour, such cross-ratios exist only starting from six points. Therefore, the Ward identity uniquely determines F_4 and F_5 in terms of $\Gamma_{\text{cusp}}(a)$, and a coupling-dependent constant. For example,

$$F_4\left(x_{13}^2/x_{24}^2\right) = \frac{1}{4} \Gamma_{\text{cusp}}(a) \log^2 \frac{x_{13}^2}{x_{24}^2} + C(a), \tag{4.29}$$

is the unique solution to Eq. (4.27). Plugging this into the definition of the regulator-independent quantity H_4, we obtain Eq. (4.16). This formula is expected to be correct in the planar limit, and to all loop orders, for both the Wilson loop and the scattering amplitude.

Similarly, the functional dependence of H_5 is uniquely determined by the conformal Ward identities. We will leave the solution as an exercise to the reader. This implies that the functional form of the four- and five-particle amplitudes in $\mathcal{N} = 4$ super Yang-Mills are known to all orders in the coupling constant! Moreover, the cusp anomalous dimension $\Gamma_{\text{cusp}}(a)$ is believed to be known exactly from the hidden integrability of the theory in the large N limit [13].

A new interesting feature appears starting from $n = 6$ points/particles. In this case, the Ward identities fix the answer only up to an a priori arbitrary function of conformal invariants. The latter take the general form of cross-ratios

$$u_{ijkl} = \frac{x_{ij} x_{kl}}{x_{ik}^2 x_{jl}^2}. \tag{4.30}$$

Indeed, it is easy to see that $K^\mu u_{ijkl} = 0$. These cross-ratios usually appear already at $n = 4$, but in the present case their appearance is delayed to $n = 6$ due to the null conditions $x^2_{i,i+1}$. At $n = 6$ points, there are three independent cross-ratios

$$u = \frac{x^2_{13}x^2_{46}}{x^2_{14}x^2_{36}}, \qquad v = \frac{x^2_{24}x^2_{15}}{x^2_{25}x^2_{14}}, \qquad w = \frac{x^2_{35}x^2_{26}}{x^2_{36}x^2_{25}}. \qquad (4.31)$$

Therefore, given a particular solution F_6^{part} (see exercise) to the Ward identity (4.27), the general solution can be written as

$$F_6 = F_6^{\text{part}} + R(u, v, w; a). \qquad (4.32)$$

The so-called remainder function $R(u, v, w; a)$ is currently under intense investigation and a lot is known about it. It is remarkable that six-particle amplitudes can be reduced to a function of only three variables.

Similar restrictions are valid for non-MHV amplitudes. There, one defines a so-called ratio function [19] $\mathscr{R}$

$$\mathscr{A}_6^{\text{non-MHV}} = \mathscr{A}_6^{\text{MHV}} \times e^{\mathscr{R}}, \qquad (4.33)$$

where $\mathscr{A}_6^{\text{MHV}}$ is the complete, i.e. loop-level MHV amplitude. Since the latter gives a particular solution to the dual conformal anomaly equation, $\mathscr{R}$ is expected to be a function of conformal invariants only (in the limit $\varepsilon \to 0$).

The conformal symmetry of the Wilson loops acts linearly on the variables x_i^μ. From the point of view of scattering amplitudes, these are dual variables, since $x_{i+1}^\mu - x_i^\mu = p_i^\mu$. The Wilson loop/scattering amplitude duality therefore predicts that scattering amplitudes should have a (broken) dual conformal symmetry. How this symmetry is realized will be discussed in the next subsection.

Exercise 4.3 (Consequences of Conformal Ward Identity)

(a) Verify that the formula given in Eq. (4.29) is the most general solution to the Ward identity given in Eq. (4.27) for $n = 4$ points.
(b) Determine the most general solution of Eq. (4.27) for $n = 5$ and $n = 6$ points.

4.4.2 Dual Conformal Symmetry of Scattering Amplitudes

In the previous section we saw that the loop corrections to MHV scattering amplitudes should have a broken dual conformal symmetry, just like the dual Wilson loops. In fact, hints for this symmetry can be seen at the level of the planar loop integrand. It was observed that, ignoring regularization, the planar loop integrand is covariant under dual conformal transformations [20].

As an example, consider the one-loop corrections to the four-particle amplitude, computed in dimensional regularization. As usual for MHV amplitudes, it is convenient to factor out the tree-level contribution,

$$A_4 = A_4^{\text{tree}} \times M_4. \qquad (4.34)$$

Omitting the overall normalization the one-loop amplitude is given by the following one-loop box integral, as was shown in Chap. 3, cf. Eq. (3.108)

$$
\begin{aligned}
M_4^{(1)} &= \int \frac{d^D k}{i\pi^{D/2}} \frac{st}{(-k^2)(-(k+p_1)^2)(-(k+p_1+p_2)^2)(-(k+p_1+p_2+p_3)^2)} \\
&= \int \frac{d^D x_0}{i\pi^{D/2}} \frac{x_{13}^2 x_{24}^2}{x_{10}^2 x_{20}^2 x_{30}^2 x_{40}^2}.
\end{aligned}
\tag{4.35}
$$

In order to study the symmetries, which are broken by the $D = 4 - 2\varepsilon$ dimensional loop integration, it is useful to define the *integrand*

$$
I(x_1, x_2, x_3, x_4; x_0) := \frac{x_{13}^2 x_{24}^2}{x_{10}^2 x_{20}^2 x_{30}^2 x_{40}^2}.
\tag{4.36}
$$

It is easy to see that it transforms covariantly under special conformal transformations. In order to see this, recall that special conformal transformations can be implemented by a sequence of translation, inversion, and another translation, cf. Exercise 2.7. Since I is translation invariant, we only need to verify its behavior under inversions $x^\mu \to x^\mu / x^2$. Using the fact that $x_{ij}^2 \to x_{ij}^2 / (x_i^2 x_j^2)$, we immediately see that

$$
I(x_1, x_2, x_3, x_4; x_0) \overset{x^\mu \to x^\mu / x^2}{\to} \left(x_0^2\right)^4 I(x_1, x_2, x_3, x_4; x_0).
\tag{4.37}
$$

The covariance factor $(x_0^2)^4$ is precisely cancelled by the Jacobian of a four dimensional loop integration. If the integral were finite in four dimensions, it would have an exact dual conformal symmetry. Due to infrared divergences, this symmetry is broken, just as for the dual Wilson loops, but in a controlled way.

Dual conformal symmetry is very helpful in constraining the loop integrand, because it puts a constraint on the number of numerators/propagators for each loop. The reason that this constraint works so well at the level of the loop integrand is probably related to a natural regularization that preserves this symmetry [21].

Working in the dual coordinate space, it is easy to write down all dual conformal integrals at a given loop order. Examples up to three loops are given in Fig. 4.6. At higher loops, certain naively dual conformal integrals do not appear due to unphysical singularities. In practice, dual conformal symmetry is extremely useful when making an ansatz for loop integrands, especially in combination with (generalized) unitarity methods. For example, this proved to be a useful guiding principle when computing the four-loop amplitude in [22]. In Ref. [23] it was shown that in combination with certain special cuts, dual conformal symmetry uniquely determines the four-point integrand, and that this is a very useful approach in practice.

The example above was for the loop corrections to the four-particle MHV amplitude. This is historically how the dual conformal symmetry was first observed for scattering amplitudes. The separation between the tree-level and the loop-level structure is special to MHV amplitudes. In order to understand the symmetry for non-MHV amplitudes, it is important to understand how it acts also on the tree-level

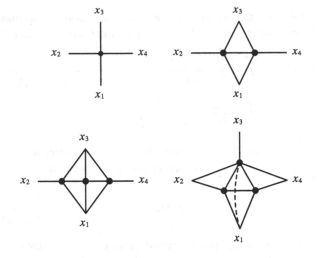

Fig. 4.6 Dual conformal integrals that appear in four-point scattering amplitudes at one and two loops (*first row*) and three loops (*second row*). The *dashed line* indicates a numerator factor, the *black dots* stand for dual x-space integrations. Numerator factors of x_{13}^2 or x_{24}^2 are not displayed

factor. In order to do this, one has to include supersymmetry, and generalize from a dual conformal to a dual superconformal symmetry group. This will be discussed in the next section.

4.5 Dual Superconformal and Yangian Symmetry

We have seen previously how dual conformal symmetry acts on the bosonic Wilson loop dual to MHV scattering amplitudes, in the space of dual coordinates. In order to be able to discuss the generic amplitudes, the action of the dual generators on λ, $\tilde{\lambda}$ has to be defined. This was done in Ref. [19]. For reasons of brevity, we will not repeat the explicit construction and form of the generators here. What is important is that with these definitions, the objects entering tree-level scattering amplitudes, such as $\langle i(i+1)\rangle$, $\langle i|x_{ij}x_{jk}|k\rangle$, etc., transform covariantly under the finite transformations. In this way, it is simple to verify the dual superconformal symmetry of the expressions appearing in tree-level scattering amplitudes, just as for loop integrands, where we only needed the bosonic part of the transformations.

Moreover, it is also possible to give an infinitesimal form of the dual superconformal generators under which the tree-level amplitudes are invariant. Perhaps the most natural way of writing them down is on the extended superspace $\{\lambda^\alpha, \tilde{\lambda}^{\dot\alpha}, \eta^A, x^{\alpha\dot\alpha}, \theta^{\alpha A}\}$, subject to the constraints

$$\lambda_i^\alpha \tilde{\lambda}_i^{\dot\alpha} = x_i^{\alpha\dot\alpha} - x_{i+1}^{\alpha\dot\alpha}, \qquad \lambda_i^\alpha \eta_i^A = \theta_i^{\alpha A} - \theta_{i+1}^{\alpha A}. \tag{4.38}$$

The set of generators is $\{P, K, S, \bar{S}, Q, \bar{Q}\}$, dual Lorentz generators and dual $SU(4)$ rotations. Explicitly one has for example the dual superconformal K and S generators

$$K^{\alpha\dot\alpha} = \sum_{i=1}^{n} \left[x_i^{\alpha\dot\beta} x_i^{\dot\alpha\beta} \frac{\partial}{\partial x_i^{\beta\dot\beta}} + x_i^{\dot\alpha\beta} \theta_i^{\alpha B} \frac{\partial}{\partial \theta_i^{\beta B}} + x_i^{\dot\alpha\beta} \lambda_i^{\alpha} \frac{\partial}{\partial \lambda_i^{\beta}} \right.$$

$$\left. + x_{i+1}^{\alpha\dot\beta} \tilde\lambda_i^{\dot\alpha} \frac{\partial}{\partial \tilde\lambda_i^{\dot\beta}} + \tilde\lambda_i^{\dot\alpha} \theta_{i+1}^{\alpha B} \frac{\partial}{\partial \eta_i^{B}} \right], \tag{4.39}$$

$$S_\alpha^A = \sum_{i=1}^{n} \left[-\theta_{i\alpha}^{B} \theta_i^{\beta A} \frac{\partial}{\partial \theta_i^{\beta B}} + x_{i\alpha}^{\dot\beta} \theta_i^{\beta A} \frac{\partial}{\partial x_i^{\beta\dot\beta}} + \lambda_{i\alpha} \theta_i^{\gamma A} \frac{\partial}{\partial \lambda_i^{\gamma}} \right.$$

$$\left. + x_{i+1\alpha}^{\dot\beta} \eta_i^A \frac{\partial}{\partial \tilde\lambda_i^{\dot\beta}} - \theta_{i+1\alpha}^{B} \eta_i^A \frac{\partial}{\partial \eta_i^{B}} \right], \tag{4.40}$$

acting in the extended superspace $\{\lambda^\alpha, \tilde\lambda^{\dot\alpha}, \eta^A, x^{\alpha\dot\alpha}, \theta^{\alpha A}\}$. They may be seen to commute with the constraints Eq. (4.38).

We have seen that scattering amplitudes in $\mathcal{N} = 4$ super Yang-Mills possess two sets of symmetries, (super)conformal symmetry thanks to the invariance of the Lagrangian, and dual (super)conformal symmetry thanks to the duality with Wilson loops. A natural question then is what the closure of the two symmetry algebras is. This question has been answered in [24] by first choosing a realization of the symmetry generators of both algebras on the same space of variables, and then analyzing their commutation relations.

In order to compare the dual superconformal generators with the generators of the original superconformal group, it is useful to solve the constraints (4.38) and express the generators in terms of $\{\lambda^\alpha, \tilde\lambda^{\dot\alpha}, \eta^A\}$ only. This is done via

$$x_i^{\alpha\dot\alpha} = x_1^{\alpha\dot\alpha} - \sum_{j<i} \lambda_j^\alpha \tilde\lambda_j^{\dot\alpha}, \qquad \theta_i^{\alpha A} = \theta_1^{\alpha A} - \sum_{j<i} \lambda_j^\alpha \eta_j^A \quad \text{for } 2 \le i \le n+1. \tag{4.41}$$

Doing this, one sees that some of the generators become trivial, namely P and Q, while others overlap with the original superconformal generators, namely $\bar S$ and $\bar Q$, see Fig. 4.7. (Generators representing rotations and charges are not shown in the figure.) However, there are also new generators, namely K and S. In fact, in view of the commutation relations of the algebra, it is sufficient to consider only one new non-trivial operators, S. Up to pieces ΔS that trivially annihilate the scattering amplitudes, it is explicitly given by

$$S_\alpha^A + \Delta S_\alpha^A = \frac{1}{2} \sum_{i>j} \left[m_{i\alpha}^\gamma q_{j\gamma}^A - \frac{1}{2}(d_i + c_i) q_{j\alpha}^A + p_{i\alpha}^{\dot\beta} \bar s_{j\dot\beta}^A + q_{i\alpha}^B r_{jB}^A - (i \leftrightarrow j) \right]. \tag{4.42}$$

This is a symmetry of tree-level scattering amplitudes.

In summary, we see that there is a certain overlap between the two algebras. The algebraic structure one finds is that of a Yangian algebra of $psu(2,2|4)$, denoted $Y(psu(2,2|4))$. The Yangian algebra $Y(\mathfrak{g})$ of a simple Lie algebra $\mathfrak{g}$ was introduced

Fig. 4.7 Relation between
superconformal and dual
superconformal algebras

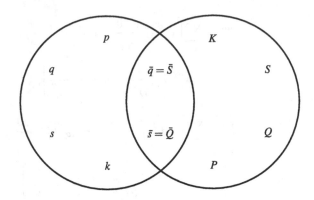

by Drinfeld [25, 26]. It is a deformation of the loop-algebra spanned by the generators $J_a^{(n)}$ with grading $n \in \mathbb{N}$. One demands the level-zero and level-one commutation relations

$$\left[J_a^{(0)}, J_b^{(0)}\right\} = f_{ab}^c J_c^{(0)}, \qquad \left[J_a^{(0)}, J_b^{(1)}\right\} = f_{ab}^c J_c^{(1)}, \tag{4.43}$$

where we use mixed brackets $[.,.\}$ to denote the graded commutator.[1] In fact the first
two level generators $J_a^{(0)}$ and $J_a^{(1)}$ span all of $Y(\mathfrak{g})$. In addition the representations
of $J_a^{(0)}$ and $J_a^{(1)}$ need to obey a generalized Jacobi identity, known as Serre relations.
The higher level generators follow from commutators of the level-one generators.
The distinguishing feature of the Yangian is a non-trivial co-product for the level-
one generators

$$\Delta\left(J_a^{(0)}\right) = J_a^{(0)} \otimes \mathbf{1} + \mathbf{1} \otimes J_a^{(0)}, \tag{4.44}$$

$$\Delta\left(J_a^{(1)}\right) = J_a^{(1)} \otimes \mathbf{1} + \mathbf{1} \otimes J_a^{(1)} + f_a{}^{bc} J_b^{(0)} \otimes J_c^{(0)}. \tag{4.45}$$

Note that in the last term quadratic in $J_a^{(0)}$ the structure constant with 'inverted'
indices appears. Indices are raised and lowered with the group metric $\text{Tr}(J_{Ra}^{(0)} J_{Rb}^{(0)})$
with $J_{R,a}^{(0)}$ in the defining representation of $\mathfrak{g}$.

The level zero generators $J_a^{(0)}$ of the Yangian as it appears in the superamplitude
problem coincide with the generators of $psu(2,2|4)$. Moreover, one finds that the
level one generators acting on all legs of the amplitude are explicitly given by

$$J_a^{(1)} = f_a{}^{cb} \sum_{i \leq 1 < j \leq n} J_{ib}^{(0)} J_{jc}^{(0)}. \tag{4.46}$$

This implies a vanishing of the first two terms of Eq. (4.45) when acting on the
amplitude $\mathscr{A}$, i.e. a trivial evaluation representation. We see that up to a convention

---

[1] I.e. $[O_1, O_2\} := O_1 O_2 - (-1)^{\deg(O_1)\deg(O_2)} O_2 O_1$, and where $\deg(O)$ is the Grassmann degree
of O.

dependent factor of 2, $J_a^{(1)}$ for a corresponding to q_α^A is given by Eq. (4.42) in the above. In this way, via Eq. (4.43), all other level one generators of the Yangian are generated.

The Yangian is a symmetry of tree-level amplitude, i.e. for any generator $J \in Y(psu(2,2|4))$, we have

$$J \mathscr{A} = 0, \tag{4.47}$$

up to contact terms, which are related to infrared divergences [27]. Similarly, as we have already seen in part in the previous section, the Yangian symmetry also strongly constrains the structure of planar loop integrands. At the integrated level, the symmetry is broken by infrared divergences. See [28–30] for more details.

The appearance of a Yangian as the underlying symmetry algebra of scattering amplitudes in $\mathcal{N} = 4$ SYM is very exciting, as it is often the hallmark for integrability. It typically arises in *two-dimensional* integrable quantum field theories, here we observe it for an interacting four-dimensional theory.

We have already seen in the section on tree-level amplitudes that choosing appropriate variables is very helpful for solving problems. With hindsight we see that the reason why the answer in dual variables was so simple is indeed the underlying symmetry. One can introduce even better (super-)twistor variables (see [8, 9, 24]). They are related to the chiral superspace variables $\{\lambda, \tilde{\lambda}, \eta\}$ via a Fourier transformation in the λ_i^α to $\mu_{i\alpha}$ leading us to $\mathscr{Z}_i^{\bar{A}} = \{\mu^\alpha, \tilde{\lambda}^{\dot{\alpha}}, \eta^A\}$. Expressed in the these variables the Yangian generators take a particularly simple form. Explicitly the level zero and one generators take the compact form,

$$J^{(0)\bar{A}}{}_{\bar{B}} = \sum_i \mathscr{Z}_i^{\bar{A}} \frac{\partial}{\partial \mathscr{Z}_i^{\bar{B}}}, \tag{4.48}$$

and

$$J^{(1)\bar{A}}{}_{\bar{B}} = \sum_{i>j} \left[\mathscr{Z}_i^{\bar{A}} \mathscr{Z}_j^{\bar{C}} \frac{\partial}{\partial \mathscr{Z}_i^{\bar{C}}} \frac{\partial}{\partial \mathscr{Z}_j^{\bar{B}}} - (i \leftrightarrow j) \right]. \tag{4.49}$$

With this form of the symmetry generators, the Yangian symmetry can be seen easily in the recent Grassmannian formulation of leading singularities of scattering amplitudes [31].

4.6 From Correlation Functions to Wilson Loops and Scattering Amplitudes

In quantum field theories, a natural object apart from scattering amplitudes and Wilson loops are correlation functions of local gauge invariant operators $\mathscr{O}(x)$,

$$G_n = \langle \mathscr{O}(x_1) \cdots \mathscr{O}(x_n) \rangle. \tag{4.50}$$

For example, in a model with scalars ϕ and gluons, one could take $\mathscr{O}(x) =$ $\mathrm{Tr}[\phi(x)\phi(x)]$. In a conformal field theory, such as $\mathscr{N} = 4$ super Yang-Mills, G_n of Eq. (4.50) can only depend on conformal cross-ratios. Such correlation functions are interesting objects and are well-studied.

It was shown recently that they are directly related to Wilson loops [32] in a special limit. As a consequence, in $\mathscr{N} = 4$ super Yang-Mills, they are also related to scattering amplitudes, thanks to the duality between Wilson loops and scattering amplitudes reviewed in Sect. 4.3.

The correlation functions G_n have singularities when two operators come close to each other, $x_i \to x_j$. The details of this limit depend on whether we perform it in the Euclidean or Minkowskian sense. In the former case, the limit is governed by the Euclidean operator product expansion and can be expressed in terms of local operators. The Minkowskian case, where the two-points are light-like separated $(x_i - x_j)^2 = 0$ but $x_i \neq x_j$ is more interesting. It is governed by the light-cone operator product expansion. In this case the leading behavior is governed by a Wilson line connecting the two points that are light-like separated.

As shown in Ref. [32], the polygonal Wilson lines studied in Sect. 4.3 govern the leading behavior of the correlation functions G_n in a special limit where all neighboring points x_i, x_{i+1} are taken to be light-like separated, i.e. $x_{i,i+1}^2 \to 0$ (with the identification $x_{i+n} \equiv x_i$). More precisely, one has [32]

$$\lim_{x_{i,i+1}^2 \to 0} G_n / G_n^{\text{tree}} = W^{\text{adj}}[C_n], \tag{4.51}$$

where

$$W^{\text{adj}} = \frac{1}{N_c^2 - 1} \left\langle \mathrm{Tr}_{\text{adj}} P \exp\left(i \oint_{C_n} dx \cdot A(x) \right) \right\rangle. \tag{4.52}$$

In the large N_c limit one can relate this to Wilson loops in the fundamental representation, namely

$$W^{\text{adj}}[C_n] = \left(W[C_n] \right)^2 + \mathcal{O}\left(1/N_c^2 \right), \tag{4.53}$$

which translates in a factor of 2 when taking logarithms of such objects.

We close this section with a number of comments. When taking the on-shell limit of Eq. (4.51) relating the correlation functions and the Wilson loops, one has to decide whether to first let $\varepsilon \to 0$, or to consider it for general ε. Depending on the choice made, the details of how the limit is approached is different. Both ways of taking the limit are discussed in detail in reference [32].

Given the duality between Wilson loops and scattering amplitudes, Eq. (4.51) also implies a similar relationship between correlation functions and scattering amplitudes. There is a natural way of formulating this at the level of four-dimensional integrands, thereby avoiding most of the subtleties related to the regularization. In fact, for the correlation functions, one can define integrands via insertions of the Lagrangian. This is based on the fact that a derivative in the coupling constant leads to an insertion of a term $\sim \int d^D x \mathscr{L}(x)$ into the correlation function. In this way, at one loop, the $(n + 1)$-point function $\langle \mathscr{O}(x_1) \cdots \mathscr{O}(x_n) \mathscr{L}(x) \rangle$ defines the integrand for G_n. In taking the on-shell limit at the level of the integrand, one expects to find the integrand for the corresponding scattering amplitude [33].

4.7 References and Further Reading

Recursion relations for loop integrands were introduced in Ref. [7]. See also Ref. [34]. The duality between Wilson loops and scattering amplitudes was reviewed in [10, 11]. The duality was discussed in twistor space in [35]. The supersymmetric generalization was discussed in [15, 36]. The duality lead to further progress in the understanding of scattering amplitudes, as the former can be systematically expanded in the OPE limit of Wilson loops [37, 38]. Moreover, the (broken) symmetries of Wilson loops were exploited to obtain results at high loop orders [29, 30]. The relationship between correlation functions and Wilson loops was discussed in Ref. [32]. Scattering amplitudes in $\mathscr{N} = 4$ super Yang-Mills can also be discussed at strong coupling [14], thanks to the conjectured AdS/CFT correspondence. The Yangian symmetry discovered in [24] also plays an important role in the description of on-shell diagrams in $\mathscr{N} = 4$ super Yang-Mills [39, 40]. Yangian symmetry was also recently observed in (super)-Wilson loops of $\mathscr{N} = 4$ super Yang-Mills [41].

There are a number of further topics that we have not discussed here, keeping in mind that these lecture notes should be suitable for a one-term specialized class. Some of the more recent topics include the following: color-kinematics duality [42], recent work on supergravity amplitudes [43], the formulation of on-shell diagrams in terms of Grassmannians and their relation to permutations [39]. Most of these topics are reviewed in the recent special issue of J. Phys. A [44–57], as well as the very recent comprehensive review [58].

References

1. J. Kublbeck, M. Bohm, A. Denner, FeynArts: computer algebraic generation of Feynman graphs and amplitudes. Comput. Phys. Commun. **60**, 165 (1990)
2. P. Nogueira, Automatic Feynman graph generation. J. Comput. Phys. **105**, 279 (1993)
3. T. Hahn, Generating Feynman diagrams and amplitudes with FeynArts 3. Comput. Phys. Commun. **140**, 418 (2001). arXiv:hep-ph/0012260
4. C.F. Berger, Z. Bern, L.J. Dixon, D. Forde, D.A. Kosower, Bootstrapping one-loop QCD amplitudes with general helicities. Phys. Rev. D **74**, 036009 (2006). arXiv:hep-ph/0604195
5. S. Caron-Huot, Loops and trees. J. High Energy Phys. **1105**, 080 (2011). arXiv:1007.3224
6. R.K. Ellis, W.T. Giele, Z. Kunszt, K. Melnikov, Masses, fermions and generalized D-dimensional unitarity. Nucl. Phys. B **822**, 270 (2009). arXiv:0806.3467
7. N. Arkani-Hamed, J.L. Bourjaily, F. Cachazo, S. Caron-Huot, J. Trnka, The all-loop integrand for scattering amplitudes in planar $N = 4$ SYM. J. High Energy Phys. **1101**, 041 (2011). arXiv:1008.2958
8. A. Hodges, Eliminating spurious poles from gauge-theoretic amplitudes (2009). arXiv: 0905.1473

9. L. Mason, D. Skinner, Dual superconformal invariance, momentum twistors and Grassmannians. J. High Energy Phys. **11**, 045 (2009). arXiv:0909.0250
10. L.F. Alday, R. Roiban, Scattering amplitudes, Wilson loops and the string/gauge theory correspondence. Phys. Rep. **468**, 153 (2008). arXiv:0807.1889
11. J.M. Henn, Duality between Wilson loops and gluon amplitudes. Fortschr. Phys. **57**, 729 (2009). arXiv:0903.0522
12. I.A. Korchemskaya, G.P. Korchemsky, On lightlike Wilson loops. Phys. Lett. B **287**, 169 (1992)
13. N. Beisert, B. Eden, M. Staudacher, Transcendentality and crossing. J. Stat. Mech. **07**, 01021 (2007). arXiv:hep-th/0610251
14. L.F. Alday, J.M. Maldacena, Gluon scattering amplitudes at strong coupling. J. High Energy Phys. **06**, 064 (2007). arXiv:0705.0303
15. S. Caron-Huot, Notes on the scattering amplitude/Wilson loop duality. J. High Energy Phys. **1107**, 058 (2011). arXiv:1010.1167
16. T. Adamo, M. Bullimore, L. Mason, D. Skinner, A proof of the supersymmetric correlation function/Wilson loop correspondence. J. High Energy Phys. **1108**, 076 (2011). arXiv:1103.4119
17. J.M. Drummond, G.P. Korchemsky, E. Sokatchev, Conformal properties of four-gluon planar amplitudes and Wilson loops. Nucl. Phys. B **795**, 385 (2008). arXiv:0707.0243
18. J.M. Drummond, J. Henn, G.P. Korchemsky, E. Sokatchev, Conformal Ward identities for Wilson loops and a test of the duality with gluon amplitudes. Nucl. Phys. B **826**, 337 (2010). arXiv:0712.1223
19. J.M. Drummond, J. Henn, G.P. Korchemsky, E. Sokatchev, Dual superconformal symmetry of scattering amplitudes in $\mathcal{N} = 4$ super-Yang–Mills theory. Nucl. Phys. B **828**, 317 (2010). arXiv:0807.1095
20. J.M. Drummond, J. Henn, V.A. Smirnov, E. Sokatchev, Magic identities for conformal four-point integrals. J. High Energy Phys. **01**, 064 (2007). arXiv:hep-th/0607160
21. L.F. Alday, J.M. Henn, J. Plefka, T. Schuster, Scattering into the fifth dimension of $\mathcal{N} = 4$ super Yang-Mills. J. High Energy Phys. **01**, 077 (2010). arXiv:0908.0684
22. Z. Bern, M. Czakon, L.J. Dixon, D.A. Kosower, V.A. Smirnov, The four-loop planar amplitude and cusp anomalous dimension in maximally supersymmetric Yang-Mills theory. Phys. Rev. D **75**, 085010 (2007). arXiv:hep-th/0610248
23. J.L. Bourjaily, A. DiRe, A. Shaikh, M. Spradlin, A. Volovich, The soft-collinear bootstrap: $N = 4$ Yang-Mills amplitudes at six and seven loops. J. High Energy Phys. **1203**, 032 (2012). arXiv:1112.6432
24. J.M. Drummond, J.M. Henn, J. Plefka, Yangian symmetry of scattering amplitudes in $\mathcal{N} = 4$ super Yang-Mills theory. J. High Energy Phys. **05**, 046 (2009)
25. V.G. Drinfel'd, Hopf algebras and the quantum Yang-Baxter equation. Sov. Math. Dokl. **32**, 254 (1985)
26. V.G. Drinfel'd, Quantum groups. J. Math. Sci. **41**, 898 (1988)
27. T. Bargheer, N. Beisert, W. Galleas, F. Loebbert, T. McLoughlin, Exacting $\mathcal{N} = 4$ superconformal symmetry. J. High Energy Phys. **11**, 056 (2009). arXiv:0905.3738
28. N. Beisert, J. Henn, T. McLoughlin, J. Plefka, One-loop superconformal and Yangian symmetries of scattering amplitudes in $N = 4$ super Yang-Mills. J. High Energy Phys. **1004**, 085 (2010). arXiv:1002.1733
29. S. Caron-Huot, S. He, Jumpstarting the all-loop S-matrix of planar $N = 4$ super Yang-Mills. J. High Energy Phys. **1207**, 174 (2012). arXiv:1112.1060
30. M. Bullimore, D. Skinner, Descent equations for superamplitudes (2011). arXiv:1112.1056
31. N. Arkani-Hamed, F. Cachazo, C. Cheung, J. Kaplan, A duality for the S matrix. J. High Energy Phys. **03**, 020 (2010)
32. L.F. Alday, B. Eden, G.P. Korchemsky, J. Maldacena, E. Sokatchev, From correlation functions to Wilson loops. J. High Energy Phys. **1109**, 123 (2011). arXiv:1007.3243
33. B. Eden, G.P. Korchemsky, E. Sokatchev, From correlation functions to scattering amplitudes. J. High Energy Phys. **1112**, 002 (2011). arXiv:1007.3246

34. R.H. Boels, On BCFW shifts of integrands and integrals. J. High Energy Phys. **1011**, 113 (2010). arXiv:1008.3101
35. L.J. Mason, D. Skinner, The complete planar S-matrix of $N = 4$ SYM as a Wilson loop in twistor space. J. High Energy Phys. **1012**, 018 (2010). arXiv:1009.2225
36. S. Caron-Huot, Superconformal symmetry and two-loop amplitudes in planar $N = 4$ super Yang-Mills. J. High Energy Phys. **1112**, 066 (2011). arXiv:1105.5606
37. L.F. Alday, D. Gaiotto, J. Maldacena, A. Sever, P. Vieira, An operator product expansion for polygonal null Wilson loops. J. High Energy Phys. **1104**, 088 (2011). arXiv:1006.2788
38. B. Basso, A. Sever, P. Vieira, Space-time S-matrix and flux tube S-matrix II. Extr. Match. Data (2013). arXiv:1306.2058
39. N. Arkani-Hamed, J.L. Bourjaily, F. Cachazo, A.B. Goncharov, A. Postnikov, et al., Scattering amplitudes and the positive Grassmannian (2012). arXiv:1212.5605
40. L. Ferro, T. Lukowski, C. Meneghelli, J. Plefka, M. Staudacher, Harmonic R-matrices for scattering amplitudes and spectral regularization. Phys. Rev. Lett. **110**, 121602 (2013). arXiv:1212.0850
41. D. Muller, H. Munkler, J. Plefka, J. Pollok, K. Zarembo, Yangian symmetry of smooth Wilson loops in $N = 4$ super Yang-Mills theory. J. High Energy Phys. **1311**, 81 (2013). arXiv:1309.1676
42. Z. Bern, J.J.M. Carrasco, H. Johansson, New relations for gauge-theory amplitudes. Phys. Rev. D **78**, 085011 (2008). arXiv:0805.3993
43. Z. Bern, J.J. Carrasco, L.J. Dixon, H. Johansson, R. Roiban, The ultraviolet behavior of $N = 8$ supergravity at four loops. Phys. Rev. Lett. **103**, 081301 (2009). arXiv:0905.2326
44. M.S. Radu Roiban, A. Volovich, Scattering amplitudes in gauge theories: progress and outlook. J. Phys. A **44**, 450301 (2011)
45. L.J. Dixon, Scattering amplitudes: the most perfect microscopic structures in the universe. J. Phys. A **44**, 454001 (2011). arXiv:1105.0771
46. A. Brandhuber, B. Spence, G. Travaglini, Tree-level formalism. J. Phys. A **44**, 454002 (2011). arXiv:1103.3477
47. Z. Bern, Y.-t. Huang, Basics of generalized unitarity. J. Phys. A **44**, 454003 (2011). arXiv:1103.1869
48. J.J.M. Carrasco, H. Johansson, Generic multiloop methods and application to $N = 4$ super-Yang-Mills. J. Phys. A **44**, 454004 (2011). arXiv:1103.3298
49. H. Ita, Susy theories and QCD: numerical approaches. J. Phys. A **44**, 454005 (2011). arXiv:1109.6527
50. R. Britto, Loop amplitudes in gauge theories: modern analytic approaches. J. Phys. A **44**, 454006 (2011). arXiv:1012.4493
51. R.M. Schabinger, One-loop $N = 4$ super Yang-Mills scattering amplitudes in d dimensions, relation to open strings and polygonal Wilson loops. J. Phys. A **44**, 454007 (2011). arXiv:1104.3873
52. T. Adamo, M. Bullimore, L. Mason, D. Skinner, Scattering amplitudes and Wilson loops in twistor space. J. Phys. A **44**, 454008 (2011). arXiv:1104.2890
53. H. Elvang, D.Z. Freedman, M. Kiermaier, SUSY ward identities, superamplitudes, and counterterms. J. Phys. A **44**, 454009 (2011). arXiv:1012.3401
54. J.M. Drummond, Tree-level amplitudes and dual superconformal symmetry. J. Phys. A **44**, 454010 (2011). arXiv:1107.4544
55. J.M. Henn, Dual conformal symmetry at loop level: massive regularization. J. Phys. A **44**, 454011 (2011). arXiv:1103.1016
56. T. Bargheer, N. Beisert, F. Loebbert, Exact superconformal and Yangian symmetry of scattering amplitudes. J. Phys. A **44**, 454012 (2011). arXiv:1104.0700
57. J. Bartels, L.N. Lipatov, A. Prygarin, Integrable spin chains and scattering amplitudes. J. Phys. A **44**, 454013 (2011). arXiv:1104.0816
58. H. Elvang, Y.-t. Huang, Scattering amplitudes (2013). arXiv:1308.1697

Appendix A
Renormalization Properties of Wilson Loops

Wilson lines and loops are very fundamental objects in gauge theories. A Wilson line $W[C_{x_1,x_2}]$ depends on a contour C having endpoints x_1 and x_2. It is explicitly given by

$$W[C] = \frac{1}{N} P e^{ig \int_C ds\dot{x}^\mu A_\mu(x(s))},\tag{A.1}$$

where P stands for path ordering along the contour C, and $A_\mu = A_\mu^a T^a$, and T^a are the generators of the gauge group. For definiteness, we are going to take the gauge group to be $SU(N)$, the generators to be in the fundamental, normalized by $\text{Tr}(T^a T^b) = \delta^{ab}$.

By definition, under gauge transformations the Wilson line transforms covariantly at the endpoints. It can therefore be used for example to make a non-local configuration of elementary fields sitting at points x_1 and x_2 gauge invariant. Such non-local objects can be used, for example, as generating functions of operators with spin, as one lets x_2 approach x_1 in a light-like direction. One can also form gauge invariant non-local quantities from the Wilson line itself, by identifying the two endpoints, and taking a trace, thereby defining a Wilson loop, which is a functional of the contour C. This is the case we will discuss from now on. It turns out that Wilson loops having specific contours describe many interesting physical effects. For example, certain configurations of Wilson lines/loops describe the infrared physics of (massive) scattering amplitudes [1, 2].

In general, in a quantum field theory we have to regularize and renormalize the wave functions of the elementary fields and the coupling constant. Once this has been carried out, there can be additional ultraviolet divergences coming from specific operators introduced into the theory, for example for composite operators. This is well understood and gives rise to a renormalization of the composite operators, described by renormalization group equations, operator mixing, and anomalous dimensions. We refer the reader to [3] for a general treatment, and to [4] for examples in the context of conformal field theory.

A natural question is what the renormalization properties of the non-local Wilson loop operators defined above are, depending on the shape of the contour C. In

J.M. Henn, J.C. Plefka, *Scattering Amplitudes in Gauge Theories*,
Lecture Notes in Physics 883, DOI 10.1007/978-3-642-54022-6,
© Springer-Verlag Berlin Heidelberg 2014

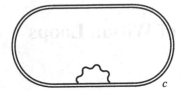 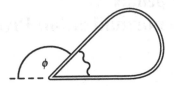

Fig. A.1 Wilson loops defined on a smooth contour (*left*) and a contour with a cusp (*right*). In the first case, there are only divergences related to mass renormalization, in the second case there are UV divergences due to the cusp in the contour

other words, we will be interested in the renormalization properties of the vacuum expectation value

$$\langle W[C]\rangle := \frac{1}{N}\mathrm{Tr}P\langle e^{ig\int_C ds\dot{x}^\mu A_\mu(x(s))}\rangle. \tag{A.2}$$

The answer turns out to be very similar to that for local operators.

There are several cases depending on the shape of the contour. We will begin by discussing smooth contours, followed by contours having one or more cusps, and self-intersections, and finally we will discuss the case where the contour contains light-like segments. More details can be found in the original papers, see e.g. [5–8].

We will assume in the following that the renormalization of the Yang-Mills theory has already been carried out, and only discuss the intrinsic divergences associated to the Wilson loop operators. In this case, Wilson loops defined in smooth, non-intersecting contours only have a linear divergence proportional to the length of the contour [6]. This divergence can be removed by a multiplicative renormalization. One can think of this as the mass renormalization of the Wilson line when it is viewed as the effective description of a heavy particle. See Fig. A.1. In dimensional regularization, such a divergence is absent, as it is powerlike.

In the case where the contour has discontinuities, there are specific divergences associated to the cusp points. See Fig. A.1. Such Wilson loops with cusps can be multiplicatively renormalized, i.e.

$$\langle W_R(C)\rangle = Z(\phi)\langle W(C)\rangle, \tag{A.3}$$

where the renormalization factor depends only locally on the contour C, through the cusp angle ϕ and on the (dimensional) regularization parameter ε, and scale μ^2, and on the Yang-Mills coupling constant. The locality of the counterterms is an important feature of ultraviolet divergences. For more than one cusp, it implies that the renormalization takes a factorized form, e.g.

$$Z(\phi_1,\ldots,\phi_n) = Z(\phi_1)\cdots Z(\phi_n). \tag{A.4}$$

From Z one can define an anomalous dimension in the usual way,

$$\Gamma(\phi, g_R) = \lim_{\varepsilon\to 0} Z\mu\frac{\partial}{\partial\mu}Z^{-1}, \tag{A.5}$$

Fig. A.2 Two Wilson loops which mix under renormalization. Loop (**a**) self-intersects, whereas (**b**) consists of two cusped loops

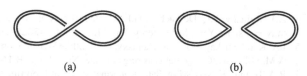

(a)   (b)

where g_R is the renormalized coupling constant. The form of Z is restricted by a renormalization group equation.

In the case of self-intersections, the analysis of UV divergences is very similar, except that one now gets nontrivial color dependence. In fact, one sees that in analogy with local operators, there are Wilson loops that mix under renormalization. For example, in the case shown in Fig. A.2, one has mixing between the two Wilson loops shown in (a) and (b). Denoting the doublet by $\mathbf{W} = (W[C_a], W[C_b])$, one now has a renormalization matrix Z,

$$\langle \mathbf{W}_R \rangle = Z(\phi) \langle \mathbf{W} \rangle. \tag{A.6}$$

If a Wilson loop contains light-like segments this leads to additional logarithmic divergences. One can think about this as a limit of the general case. For example, consider a Wilson loop having a cusp with Euclidean cusp angle $\phi = p \cdot q / \sqrt{p^2 q^2}$, where p^μ and q^μ are the momenta forming the cusp (cf. Fig. A.1). We can analytically continue to $\phi \to i\theta$, and consider the limit $\theta \to \infty$. This corresponds to the limit where one (or both) of the segments p^μ or q^μ becomes light-like. In this limit, the cusp anomalous dimension has the behavior [8]

$$\Gamma_{\text{cusp}}(\phi, g) \sim \lim_{\theta \to \infty} \theta \, \Gamma_{\text{cusp}}(g) + \mathcal{O}(1). \tag{A.7}$$

A case of physical interest will be the contour formed by n light-like segments, with cusp points $x_1, \ldots, x_n$. What is the general structure of such a correlation function? Let us choose dimensional regularization to regularize UV divergences (for renormalization of the Lagrangian, as well as for the Wilson loop). In a conformal theory with beta function $\beta = -\varepsilon g$, the renormalization group equations can be solved, leading to the simple general form [5]

$$\log \langle W \rangle = -\frac{1}{4} \sum_{\ell \geq 1} a^l \sum_{i=1}^{n} (-x_{i-1,i+1}\mu)^{\ell\varepsilon} \left(\frac{\Gamma_{\text{cusp}}^{(\ell)}}{(\ell\varepsilon)^2} + \frac{\Gamma^{(\ell)}}{\ell\varepsilon} \right), \quad a = \frac{g^2 N}{8\pi^2}. \tag{A.8}$$

References

1. G.P. Korchemsky, A.V. Radyushkin, Infrared factorization, Wilson lines and the heavy quark limit. Phys. Lett. B **279**, 359–366 (1992). arXiv:hep-ph/9203222
2. G.P. Korchemsky, A.V. Radyushkin, Loop space formalism and renormalization group for the infrared asymptotics of QCD. Phys. Lett. B **171**, 459 (1986)
3. J.C. Collins, Renormalization. An Introduction to Renormalization, the Renormalization Group, and the Operator Product Expansion (Cambridge Univ. Press, Cambridge, 1984), 390 p

4. A.V. Belitsky, S.E. Derkachov, G.P. Korchemsky, A.N. Manashov, Superconformal operators in $N = 4$ super-Yang-Mills theory. Phys. Rev. D **70**, 045021 (2004). arXiv:hep-th/0311104
5. I.A. Korchemskaya, G.P. Korchemsky, On lightlike Wilson loops. Phys. Lett. B **287**, 169 (1992)
6. A.M. Polyakov, Gauge fields as rings of glue. Nucl. Phys. B **164**, 171 (1980)
7. R.A. Brandt, F. Neri, M.-a. Sato, Renormalization of loop functions for all loops. Phys. Rev. D **24**, 879 (1981)
8. G.P. Korchemsky, A.V. Radyushkin, Renormalization of the Wilson loops beyond the leading order. Nucl. Phys. B **283**, 342 (1987)

Appendix B
Conventions and Useful Formulae

- Index and metric conventions:

$$\eta_{\mu\nu} = \mathrm{diag}(+,-,-,-), \qquad p_\mu p^\mu = p_0^2 - \mathbf{p}^2,$$

$$\varepsilon^{12} = \varepsilon^{\dot{1}\dot{2}} = \varepsilon_{21} = \varepsilon_{\dot{2}\dot{1}} = +1, \qquad \varepsilon^{21} = \varepsilon^{\dot{2}\dot{1}} = \varepsilon_{12} = \varepsilon_{\dot{1}\dot{2}} = -1,$$

$$\left(\bar{\sigma}^\mu\right)^{\dot{\alpha}\alpha} = (\mathbb{1}, -\sigma), \qquad \left(\sigma^\mu\right)_{\alpha\dot{\alpha}} = \varepsilon_{\alpha\beta}\varepsilon_{\dot{\alpha}\dot{\beta}}\left(\bar{\sigma}^\mu\right)^{\dot{\beta}\beta} = (\mathbb{1}, \sigma),$$

$$\left(\bar{\sigma}_\mu\right)^{\dot{\alpha}\alpha} = (\mathbb{1}, \sigma), \qquad \left(\sigma_\mu\right)_{\alpha\dot{\alpha}} = (\mathbb{1}, -\sigma), \qquad \text{(B.1)}$$

$$\sigma_1 = \begin{pmatrix} 0 & 1 \\ 1 & 0 \end{pmatrix}, \qquad \sigma_2 = \begin{pmatrix} 0 & -i \\ i & 0 \end{pmatrix}, \qquad \sigma_3 = \begin{pmatrix} 1 & 0 \\ 0 & -1 \end{pmatrix},$$

$$\chi^\alpha = \varepsilon^{\alpha\beta}\chi_\beta, \qquad \tilde{\chi}^{\dot{\alpha}} = \varepsilon^{\dot{\alpha}\dot{\beta}}\tilde{\chi}_{\dot{\beta}}, \qquad (\chi_\alpha)^* = \tilde{\chi}_{\dot{\alpha}},$$

$$\left(\chi^\alpha \psi^\beta\right)^\dagger = \left(\psi^\beta\right)^\dagger \left(\chi^\alpha\right)^\dagger.$$

- Spinor helicity relations:

$$p^{\alpha\dot{\alpha}} = \lambda^\alpha \tilde{\lambda}^{\dot{\alpha}}, \qquad p_{\dot{\alpha}\alpha} = \varepsilon_{\dot{\alpha}\dot{\beta}}\varepsilon_{\alpha\beta}\lambda^\beta \tilde{\lambda}^{\dot{\beta}},$$

$$\lambda_\alpha = \varepsilon_{\alpha\beta}\lambda^\beta, \qquad \tilde{\lambda}_{\dot{\alpha}} = \varepsilon_{\dot{\alpha}\dot{\beta}}\tilde{\lambda}^{\dot{\beta}},$$

$$u_+(p) = v_-(p) = \begin{pmatrix} \lambda_\alpha \\ 0 \end{pmatrix} =: |p\rangle, \qquad u_-(p) = v_+(p) = \begin{pmatrix} 0 \\ \tilde{\lambda}^{\dot{\alpha}} \end{pmatrix} =: |p], \quad \text{(B.2)}$$

$$\bar{u}_+(p) = \bar{v}_-(p) = \begin{pmatrix} 0 & \tilde{\lambda}_{\dot{\alpha}} \end{pmatrix} =: [p|, \qquad \bar{u}_-(p) = \bar{v}_+(p) = \begin{pmatrix} \lambda^\alpha & 0 \end{pmatrix} =: \langle p|,$$

$$\langle \lambda_i \lambda_j \rangle := \lambda_i^\alpha \lambda_{j\alpha}, \qquad [\tilde{\lambda}_i \tilde{\lambda}_j] := \tilde{\lambda}_{i\dot{\alpha}}\tilde{\lambda}_j^{\dot{\alpha}}$$

using the chiral representation of the Dirac matrices

$$\not{p} = p_\mu \Gamma^\mu = \begin{pmatrix} 0 & p_{\alpha\dot{\alpha}} \\ p^{\dot{\alpha}\alpha} & 0 \end{pmatrix}$$

using $p_{\alpha\dot{\alpha}} := p_\mu \left(\sigma^\mu\right)_{\alpha\dot{\alpha}}, \; p^{\dot{\alpha}\alpha} = \varepsilon^{\alpha\beta}\varepsilon^{\dot{\alpha}\dot{\beta}}p_{\beta\dot{\beta}} = p_\mu \left(\bar{\sigma}^\mu\right)^{\dot{\alpha}\alpha}.$ \quad (B.3)

J.M. Henn, J.C. Plefka, *Scattering Amplitudes in Gauge Theories*,
Lecture Notes in Physics 883, DOI 10.1007/978-3-642-54022-6,
© Springer-Verlag Berlin Heidelberg 2014

We furthermore note

$$[i|\Gamma^{\mu}|j\rangle = \langle j|\Gamma^{\mu}|i], \qquad \langle p|\Gamma^{\mu}|p] = \lambda^{\alpha}\sigma^{\mu}_{\alpha\dot{\alpha}}\tilde{\lambda}^{\dot{\alpha}} = 2p^{\mu},$$

$$[i|\Gamma^{\mu}|j\rangle\langle l|\Gamma_{\mu}|k] = 2[ik]\langle lj\rangle. \tag{B.4}$$

- The generators of the Lorentz group in the spinor representation are given by

$$\left(\sigma^{\mu\nu}\right)_{\alpha}{}^{\beta} = \frac{1}{4}\left((\sigma^{\mu})_{\alpha\dot{\alpha}}(\bar{\sigma}^{\nu})^{\dot{\alpha}\beta} - (\sigma^{\nu})_{\alpha\dot{\alpha}}(\bar{\sigma}^{\mu})^{\dot{\alpha}\beta}\right),$$

$$\left(\bar{\sigma}^{\mu\nu}\right)^{\dot{\alpha}}{}_{\dot{\beta}} = \frac{1}{4}\left((\bar{\sigma}^{\mu})^{\dot{\alpha}\alpha}(\sigma^{\nu})_{\alpha\dot{\beta}} - (\bar{\sigma}^{\nu})^{\dot{\alpha}\alpha}(\sigma^{\mu})_{\alpha\dot{\beta}}\right). \tag{B.5}$$

- Complex conjugation properties:

$$\left(\lambda^{\alpha}\right)^{*} = \tilde{\lambda}^{\dot{\alpha}}, \qquad \langle ij\rangle^{*} = \left(\lambda_{i}^{\alpha}\lambda_{j\alpha}\right)^{*} = \left(\tilde{\lambda}_{i}^{\dot{\alpha}}\tilde{\lambda}_{j\dot{\alpha}}\right)^{*} = -[ij]. \tag{B.6}$$

- Useful trace identities

$$\mathrm{Tr}[\slashed{a}\slashed{b}\slashed{c}\slashed{d}] = 4\left[(a\cdot b)(c\cdot d) - (a\cdot c)(b\cdot d) + (a\cdot d)(b\cdot c)\right],$$

$$\mathrm{Tr}[\slashed{a}\slashed{b}] = 4(a\cdot b). \tag{B.7}$$

Solutions to the Exercises

Exercise 1.1

Given $(\bar{\sigma}^\mu)^{\dot\alpha\alpha} = (\mathbb{1}, -\sigma)$ and $(\sigma^\mu)_{\alpha\dot\alpha} = \varepsilon_{\alpha\beta}\varepsilon_{\dot\alpha\dot\beta}(\bar{\sigma}^\mu)^{\dot\beta\beta}$ we have with

$$\sigma_2 = \begin{pmatrix} 0 & -i \\ i & 0 \end{pmatrix}$$

the relations

$$\varepsilon_{\alpha\beta} = \begin{pmatrix} 0 & 1 \\ -1 & 0 \end{pmatrix} = i(\sigma_2)_{\alpha\beta}, \qquad \varepsilon_{\dot\alpha\dot\beta} = \begin{pmatrix} 0 & 1 \\ -1 & 0 \end{pmatrix} = i(\sigma_2)_{\dot\alpha\dot\beta}. \qquad \text{(B.8)}$$

Then we have

$$\left(\sigma^\mu\right)_{\dot\alpha\alpha} = -\varepsilon_{\dot\alpha\dot\beta}(\bar{\sigma}^\mu)^{\dot\beta\beta}\varepsilon_{\beta\alpha} = -i^2\left(\sigma_2\bar{\sigma}^\mu\sigma_2\right)_{\dot\alpha\alpha} = \left(\sigma_2\bar{\sigma}^\mu\sigma_2\right)^T_{\alpha\dot\alpha},$$

$$\sigma_2\bar{\sigma}^\mu\sigma_2 = \sigma_2(\mathbb{1}, -\sigma)\sigma_2$$

$$= \left(\sigma_2^2, -\sigma_2\sigma_1\sigma_2, -\sigma_2\sigma_2\sigma_2, -\sigma_2\sigma_3\sigma_2\right) = (\mathbb{1}, \sigma_1, -\sigma_2, \sigma_3),$$

$$\left(\sigma_2\bar{\sigma}^\mu\sigma_2\right)^T = (\mathbb{1}, \sigma_1, \sigma_2, \sigma_3),$$

as σ_2 is antisymmetric whereas the other Pauli matrices are symmetric. We thus have $(\sigma^\mu)_{\alpha\dot\alpha} = (\mathbb{1}, \sigma)$. The relation $(\sigma_\mu)_{\alpha\dot\alpha} = (\mathbb{1}, -\sigma)$ follows trivially from $\eta_{\mu\nu} = \text{diag}(+1, -1, -1, -1)$. To prove the third relation in Exercise 1.1 $\sigma^\mu_{\alpha\dot\alpha}\sigma_{\mu\beta\dot\beta} = 2\varepsilon_{\alpha\beta}\varepsilon_{\dot\alpha\dot\beta}$ we look at the LHS of this relation for fixed values of α and β:

$$\sigma^\mu_{1\dot\alpha}\sigma_{\mu 1\dot\beta} = \sigma^0_{1\dot\alpha}\sigma^0_{1\dot\beta} - \sigma^i_{1\dot\alpha}\sigma^i_{1\dot\beta}$$

$$= \delta_{1\dot\alpha}\delta_{1\dot\beta} - \delta_{\dot\alpha 2}\delta_{\dot\beta 2} - (-i)\delta_{\dot\alpha 2}(-i)\delta_{\dot\beta 2} - \delta_{\dot\alpha 1}\delta_{\dot\beta 1} = 0,$$

$$\sigma^\mu_{2\dot\alpha}\sigma_{\mu 2\dot\beta} = \sigma^0_{2\dot\alpha}\sigma^0_{2\dot\beta} - \sigma^i_{2\dot\alpha}\sigma^i_{2\dot\beta}$$

$$= \delta_{2\dot\alpha}\delta_{2\dot\beta} - \delta_{\dot\alpha 1}\delta_{\dot\beta 1} - (i)\delta_{\dot\alpha 1}(i)\delta_{\dot\beta 1} - (-1)\delta_{\dot\alpha 2}(-1)\delta_{\dot\beta 2} = 0,$$

J.M. Henn, J.C. Plefka, *Scattering Amplitudes in Gauge Theories*,
Lecture Notes in Physics 883, DOI 10.1007/978-3-642-54022-6,
© Springer-Verlag Berlin Heidelberg 2014

$$\sigma^{\mu}_{1\dot\alpha}\sigma_{\mu 2\dot\beta} = \sigma^{0}_{1\dot\alpha}\sigma^{0}_{2\dot\beta} - \sigma^{i}_{1\dot\alpha}\sigma^{i}_{2\dot\beta}$$

$$= \delta_{1\dot\alpha}\delta_{2\dot\beta} - \delta_{\dot\alpha 2}(-1)\delta_{\dot\beta 1} - (-i)\delta_{\dot\alpha 2}(i)\delta_{\dot\beta 1} - \delta_{\dot\alpha 1}(-1)\delta_{\dot\beta 2}$$

$$= 2(\delta_{1\dot\alpha}\delta_{2\dot\beta} - \delta_{2\dot\alpha}\delta_{1\dot\beta}) = 2\varepsilon_{\dot\alpha\dot\beta}.$$

Note that the last equation implies $\sigma^{\mu}_{1\dot\alpha}\sigma_{\mu 2\dot\beta} = -2\varepsilon_{\dot\alpha\dot\beta}$ by antisymmetry in $\dot\alpha\dot\beta$. The final relation $\varepsilon^{\alpha\beta}\varepsilon^{\dot\alpha\dot\beta}\sigma^{\mu}_{\alpha\dot\alpha}\sigma^{\nu}_{\beta\dot\beta} = 2\eta^{\mu\nu}$ follows by taking the trace on the LHS

$$\mathrm{Tr}\left(-i\sigma_2\sigma^{\nu}(i\sigma_2)(\sigma^{\mu})^{T}\right) = \mathrm{Tr}\left(\sigma_2\sigma^{\nu}\sigma_2\sigma^{\mu T}\right) = \begin{cases} -\mathrm{Tr}(\sigma^{\nu}\sigma^{\mu T}) & \text{for } \nu = 1,3 \\ +\mathrm{Tr}(\sigma^{\nu}\sigma^{\mu T}) & \text{for } \nu = 0,2 \end{cases}$$

$$= \begin{cases} -2\delta^{\mu\nu} & \text{for } \nu = 1,3 \\ +2\eta^{\mu\nu} & \text{for } \nu = 0,2 \end{cases} = 2\eta^{\mu\nu}.$$

Exercise 1.2

(a) We have

$$\Gamma^{\mu}k_{\mu} = \begin{pmatrix} k^{0} & -\sigma\cdot\mathbf{k} \\ +\sigma\cdot\mathbf{k} & -k^{0} \end{pmatrix}$$

$$= \begin{pmatrix} k^{0} & 0 & -k^{3} & -\sqrt{k^{+}k^{-}}e^{-i\phi} \\ 0 & k^{0} & -\sqrt{k^{+}k^{-}}e^{+i\phi} & k^{3} \\ k^{3} & \sqrt{k^{+}k^{-}}e^{-i\phi} & -k^{0} & 0 \\ \sqrt{k^{+}k^{-}}e^{i\phi} & -k^{3} & 0 & -k^{0} \end{pmatrix},$$

which upon multiplication with $u_{+}(k)$ or $u_{-}(k)$ is easily seen to vanish. For the helicity relations we first note

$$P_{\pm} = \frac{1}{2}\begin{pmatrix} \mathbb{1} & \pm\mathbb{1} \\ \pm\mathbb{1} & \mathbb{1} \end{pmatrix}.$$

Then as $u_{\pm}(k) = \begin{pmatrix} \xi \\ \pm\xi \end{pmatrix}$ we have

$$P_{\pm}u_{\pm} = \frac{1}{2}\begin{pmatrix} \mathbb{1} & \pm\mathbb{1} \\ \pm\mathbb{1} & \mathbb{1} \end{pmatrix}\begin{pmatrix} \xi \\ \pm\xi \end{pmatrix} = \begin{pmatrix} \xi \\ \pm\xi \end{pmatrix} = u_{\pm},$$

$$P_{\pm}u_{\mp} = \frac{1}{2}\begin{pmatrix} \mathbb{1} & \pm\mathbb{1} \\ \pm\mathbb{1} & \mathbb{1} \end{pmatrix}\begin{pmatrix} \xi \\ \mp\xi \end{pmatrix} = 0.$$

(b) For an arbitrary Dirac-spinor χ we have

$$\bar\chi P_{\pm} = \chi^{\dagger}\Gamma^{0}P_{\pm} = \left(P^{\dagger}_{\pm}\Gamma^{0\dagger}\chi\right)^{\dagger} = \left(P_{\pm}\Gamma^{0}\chi\right)^{\dagger} = \left(\Gamma^{0}P_{\mp}\chi\right)^{\dagger}.$$

From this it follows that $\bar u_{\pm}P_{\mp} = \bar u_{\pm}$ and $\bar u_{\pm}P_{\pm} = 0$.

(c) Through explicit matrix multiplication one verifies that $U = \frac{1}{\sqrt{2}}(\mathbb{1} - i\Gamma^1\Gamma^2\Gamma^3) = \frac{1}{\sqrt{2}}\begin{pmatrix} 1 & -1 \\ 1 & 1 \end{pmatrix}$. It is indeed a unitary matrix. The transformed Dirac matrices then take the form

$$\Gamma^0 \to \frac{1}{2}\begin{pmatrix} 1 & -1 \\ 1 & 1 \end{pmatrix}\begin{pmatrix} 1 & 0 \\ 0 & -1 \end{pmatrix}\begin{pmatrix} 1 & 1 \\ -1 & 1 \end{pmatrix} = \begin{pmatrix} 0 & 1 \\ 1 & 0 \end{pmatrix},$$

$$\Gamma \to \frac{1}{2}\begin{pmatrix} 1 & -1 \\ 1 & 1 \end{pmatrix}\begin{pmatrix} 0 & \sigma \\ -\sigma & 0 \end{pmatrix}\begin{pmatrix} 1 & 1 \\ -1 & 1 \end{pmatrix} = \begin{pmatrix} 0 & \sigma \\ -\sigma & 0 \end{pmatrix}.$$

Thus the Dirac-matrices in the transformed basis read

$$\Gamma^\mu_{ch} = \begin{pmatrix} 0 & \sigma^\mu_{\alpha\dot\alpha} \\ \bar\sigma^{\mu\dot\alpha\alpha} & 0 \end{pmatrix} \quad \text{and} \quad \Gamma^5_{ch} = \begin{pmatrix} 1 & 0 \\ 0 & -1 \end{pmatrix}.$$

Expressing the solutions to the Dirac equation in the chiral basis it follows that

$$U u_+ = \begin{pmatrix} 0 \\ 0 \\ \sqrt{k^+} \\ \sqrt{k^-}e^{i\phi} \end{pmatrix} \quad \text{and} \quad U u_- = \begin{pmatrix} \sqrt{k^-}e^{-i\phi} \\ -\sqrt{k^+} \\ 0 \\ 0 \end{pmatrix}.$$

Note the appearance of the helicity spinors λ^α and $\tilde\lambda_\alpha$ discussed in Sect. 1.6.

Exercise 1.3

The prove of Eq. (1.93) is straightforward. With

$$[i|\Gamma^\mu|j\rangle = \tilde\lambda_{i\dot\alpha}\lambda_{j\alpha}\bar\sigma^{\mu\dot\alpha\alpha}$$

we have

$$[i|\Gamma^\mu|j\rangle[k|\Gamma^\nu|l\rangle\eta_{\mu\nu} = \tilde\lambda_{i\dot\alpha}\lambda_{j\alpha}\tilde\lambda_{k\dot\beta}\lambda_{l\beta}\underbrace{\bar\sigma^{\mu\dot\alpha\alpha}\bar\sigma^{\nu\dot\beta\beta}\eta_{\mu\nu}}_{=2\varepsilon^{\dot\alpha\dot\beta}\varepsilon^{\alpha\beta}} = 2[ik]\langle lj\rangle.$$

Exercise 1.4

We start from the full Feynman rule four-point vertex contracted with dummy polarization vectors ε_i

$$V_4 = -ig^2 f^{abe}f^{cde}(\varepsilon_1\cdot\varepsilon_3)(\varepsilon_2\cdot\varepsilon_4) - (\varepsilon_1\cdot\varepsilon_2)(\varepsilon_3\cdot\varepsilon_4) + \text{cyclic}$$

and use $f^{abe}f^{cde} = -\frac{1}{2}\text{Tr}([T^a, T^b][T^c, T^d])$ which is obtained from Eq. (1.96). Note that the $U(1)$ piece cancels out here. Expanding out the commutators in the traces and collecting terms of identical color ordering one finds

$$V_4 = \frac{ig^2}{2} \text{Tr}\left(T^a T^b T^c T^d\right)\left[2(\varepsilon_1 \cdot \varepsilon_2)(\varepsilon_3 \cdot \varepsilon_4) - (\varepsilon_1 \cdot \varepsilon_3)(\varepsilon_2 \cdot \varepsilon_4)\right.$$

$$\left. - (\varepsilon_1 \cdot \varepsilon_4)(\varepsilon_2 \cdot \varepsilon_3)\right] + \text{cyclic},$$

which is the result quoted in the color ordered Feynman rules.

Exercise 1.5

(a) Taking parity and cyclicity into account we have the independent 4-gluon amplitudes

$$A_4^{\text{tree}}(1^+, 2^+, 3^+, 4^+), \qquad A_4^{\text{tree}}(1^-, 2^+, 3^+, 4^+),$$
$$A_4^{\text{tree}}(1^-, 2^-, 3^+, 4^+), \qquad A_4^{\text{tree}}(1^-, 2^+, 3^-, 4^+).$$

The last two are related via the $U(1)$ decoupling theorem as

$$A_4^{\text{tree}}(1^-, 2^+, 3^-, 4^+) = -A_4^{\text{tree}}(1^-, 2^+, 4^+, 3^-) - A_4^{\text{tree}}(1^-, 4^+, 2^+, 3^-)$$
$$= -A_4^{\text{tree}}(3^-, 1^-, 2^+, 4^+) - A_4^{\text{tree}}(3^-, 1^-, 4^+, 2^+).$$

Hence only the three amplitudes $A_4^{\text{tree}}(1^+, 2^+, 3^+, 4^+)$, $A_4^{\text{tree}}(1^-, 2^+, 3^+, 4^+)$ and $A_4^{\text{tree}}(1^-, 2^-, 3^+, 4^+)$ are independent. In fact the first two of this list vanish, so there is only one independent 4-gluon amplitude at tree-level to be computed.

(b) Moving on two the 5-gluon trees we have the four cyclic and parity independent amplitudes

$$A_5^{\text{tree}}(1^+, 2^+, 3^+, 4^+, 5^+), \qquad A_5^{\text{tree}}(1^-, 2^+, 3^+, 4^+, 5^+),$$
$$A_5^{\text{tree}}(1^-, 2^-, 3^+, 4^+, 5^+), \qquad A_4^{\text{tree}}(1^-, 2^+, 3^-, 4^+, 5^+).$$

Looking at the following $U(1)$ decoupling relation we may again relate the last amplitude in the above list to the third one

$$A_5^{\text{tree}}(2^+, 3^-, 4^+, 5^+, 1^-)$$
$$= -A_5^{\text{tree}}(3^-, 2^+, 4^+, 5^+, 1^-) - A_5^{\text{tree}}(3^-, 4^+, 2^+, 5^-, 1^-)$$
$$\quad - A_5^{\text{tree}}(3^-, 4^+, 5^+, 2^+, 1^-)$$
$$= -A_5^{\text{tree}}(1^-, 3^-, 2^+, 4^+, 5^+) - A_5^{\text{tree}}(1^-, 3^-, 4^+, 2^+, 5^+)$$
$$\quad - A_5^{\text{tree}}(1^-, 3^-, 4^+, 5^+, 2^+).$$

Hence also for the 5-gluon case there are only three independent amplitudes: $A_5^{\text{tree}}(1^+, 2^+, 3^+, 4^+, 5^+)$, $A_5^{\text{tree}}(1^-, 2^+, 3^+, 4^+, 5^+)$ and $A_5^{\text{tree}}(1^-, 2^-, 3^+, 4^+, 5^+)$. The first two in this list vanish leaving us with one independent and non-trivial 5-gluon tree-level amplitude of MHV type.

Exercise 1.6

There are two color ordered amplitudes contributing to $A(1_{\bar{q}}^-, 2_q^+, 3^-, 4^+)$

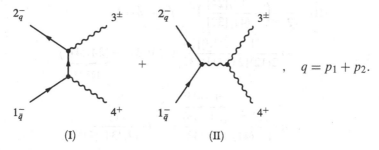

Looking at the first graph (I) we see that it is proportional to $[2|\varepsilon_3\!\!\!/\,\frac{i}{p}\varepsilon_4\!\!\!/|1\rangle$. The reference vector choice $\mu_4\tilde{\mu}_4 = p_4$ then annihilates this graph

$$(I) \sim [2|\varepsilon_3\!\!\!/\,\frac{i}{p}\varepsilon_4\!\!\!/|1\rangle = 0 \quad \text{for } \mu_4\tilde{\mu}_4 = p_4.$$

This is so as

$$\varepsilon_4\!\!\!/ = -\frac{\sqrt{2}}{\langle 4\mu_4\rangle}\big(|4]\langle\mu_4| + |\mu_4\rangle[4|\big) \quad \Rightarrow \quad \varepsilon_4\!\!\!/|1\rangle = -\frac{\sqrt{2}}{\langle 4\mu_4\rangle}|4]\langle\mu_4 1\rangle \stackrel{\mu_4=\lambda_1}{=} 0.$$

Evaluating the second graph (II) with the color-ordered Feynman rules we are led to the following expression

$$(II) = \left(-\frac{i}{\sqrt{2}}\right)^2 [2|\Gamma_\mu|1\rangle\frac{-i}{q^2}$$
$$\times \Big[\underbrace{(\varepsilon_3^- \cdot \varepsilon_4^+)(p_{34})^\mu}_{(1)} + \underbrace{(\varepsilon_4^+)^\mu(p_{4q} \cdot \varepsilon_3^-)}_{(2)} + \underbrace{(\varepsilon_3^-)^\mu(p_{q3} \cdot \varepsilon_4^+)}_{(3)}\Big], \qquad \text{(B.9)}$$

giving rise to three terms. One sees that term (2) vanishes for our choice $\mu_4\tilde{\mu}_4 = p_4$

$$(2) \sim [2|\varepsilon_4\!\!\!/^+|1\rangle \stackrel{\mu_4\tilde{\mu}_4=p_1}{=} 0.$$

For the term (3) we note $\varepsilon_3\!\!\!/^- = \frac{\sqrt{2}}{[3\mu_3]}\big(|3\rangle[\mu_3| + |\mu_3]\langle 3|\big)$ to find

$$(3) \sim [2|\varepsilon_3\!\!\!/^-|1\rangle = \frac{\sqrt{2}}{[3\mu_3]}[2\mu_3]\langle 31\rangle.$$

We now make the choice $\mu_3\tilde{\mu}_3 = p_2$ for the remaining reference vector of leg 3 which also kills this term. Hence, for these two choice of reference vectors only the term (1) in the above Eq. (B.9) contributes. One has

$$\varepsilon_3^- \cdot \varepsilon_4^+ = -\frac{\langle\mu_4 3\rangle[\mu_3 4]}{\langle 4\mu_4\rangle[3\mu_3]} \stackrel{\mu_3=\lambda_2, \mu_4=\lambda_1}{=} -\frac{\langle 13\rangle[24]}{\langle 41\rangle[32]}.$$

Inserting this into the term (1) of Eq. (B.9) and using $q^2 = \langle 12\rangle[21]$ yields

$$(\mathrm{II}) = \frac{i}{2q^2}\left(-\frac{\langle 13\rangle[24]}{\langle 41\rangle[32]}\right)[2|(\not{p}_3 - \not{p}_4)|1\rangle$$

$$= -i\frac{1}{2\langle 12\rangle[21]}\frac{\langle 13\rangle[24]}{\langle 41\rangle[32]}\big([23]\langle 31\rangle - \underbrace{[24]\langle 41\rangle}_{-[23]\langle 31\rangle}\big)$$

$$= -i\frac{\langle 13\rangle^2}{\langle 12\rangle\langle 41\rangle}\overbrace{\frac{[24]\langle 43\rangle}{[21]\langle 43\rangle}}^{-[21]\langle 13\rangle} = -i\frac{\langle 13\rangle^3}{\langle 12\rangle\langle 34\rangle\langle 41\rangle}$$

$$= -i\frac{\langle 13\rangle^3\langle 23\rangle}{\langle 12\rangle\langle 23\rangle\langle 34\rangle\langle 41\rangle},$$

as claimed. The helicity count of our result $A^{\text{tree}}_{\bar{q}qg^2} = -i\frac{\langle 13\rangle^3\langle 23\rangle}{\langle 12\rangle\langle 23\rangle\langle 34\rangle\langle 41\rangle}$ is straightforward and correct

$$h_1\big[A^{\text{tree}}_{\bar{q}qg^2}\big] = -\frac{1}{2}(3-1-1) = -\frac{1}{2}, \qquad h_2\big[A^{\text{tree}}_{\bar{q}qg^2}\big] = -\frac{1}{2}(1-1-1) = +\frac{1}{2},$$

$$h_3\big[A^{\text{tree}}_{\bar{q}qg^2}\big] = -\frac{1}{2}(4-1-1) = -1, \qquad h_4\big[A^{\text{tree}}_{\bar{q}qg^2}\big] = -\frac{1}{2}(0-1-1) = -1.$$

Exercise 2.1

   (a) We begin with the MHV case $A(1^-, 2^-, 3^+)$:

$$= \left(\frac{-gi}{\sqrt{2}}\right)\big[(\varepsilon_{-,1}\cdot\varepsilon_{-,2})(p_{12}\cdot\varepsilon_{+,3})$$

$$+ (\varepsilon_{-,2}\cdot\varepsilon_{+,3})(p_{23}\cdot\varepsilon_{-,1}) + (\varepsilon_{+,3}\cdot\varepsilon_{-,1})(p_{31}\cdot\varepsilon_{-,2})\big].$$

The gauge choice $\mu_1 = \mu_2 = \mu_3 = \mu$ annihilates the first term as then $\varepsilon_{-,1}\cdot\varepsilon_{-,2} = 0$. We note using Eq. (1.82)

$$\varepsilon_{-,2}\cdot\varepsilon_{+,3} = -\frac{\langle\mu 2\rangle[\mu 3]}{\langle 3\mu\rangle[2\mu]}, \qquad p_{23}\cdot\varepsilon_{-,1} = -\sqrt{2}\frac{[\mu 3]\langle 31\rangle}{[1\mu]}.$$

Inserting this into the amplitude (whilst droping the factor $-ig$ to go from the graph to the amplitude) yields

$$A\left(1^-,2^-,3^+\right) = -\frac{\langle\mu2\rangle[\mu3]}{\langle3\mu\rangle[2\mu]}\frac{[\mu3]\langle31\rangle}{[1\mu]} - (1\leftrightarrow2)$$

$$= \frac{[\mu3]^2}{[1u][2\mu]}\frac{\langle\mu2\rangle\langle31\rangle - \langle\mu1\rangle\langle32\rangle}{\langle3\mu\rangle}$$

$$= \frac{[\mu3]^2}{[1u][2\mu]}\langle12\rangle.$$

The MHV$_3$ kinematics implies $\tilde{\lambda}_1 \sim \tilde{\lambda}_2 \sim \tilde{\lambda}_3$ or $\tilde{\lambda}_2 = a\tilde{\lambda}_1$ and $\tilde{\lambda}_3 = b\tilde{\lambda}_1$ leading to

$$A\left(1^-,2^-,3^+\right) = \frac{b^2}{a}\langle12\rangle.$$

Momentum conservation then implies for this parametrization

$$(\lambda_1 + a\lambda_2 + b\lambda_3)\tilde{\lambda}_1 = 0,$$

which tells us that

$$a = \frac{\langle31\rangle}{\langle23\rangle} \quad \text{and} \quad b = \frac{\langle12\rangle}{\langle23\rangle}.$$

Plugging this into the above yields the final result

$$A\left(1^-,2^-,3^+\right) = \frac{\langle12\rangle^3}{\langle23\rangle\langle31\rangle}.$$

(b) The $\overline{\text{MHV}}_3$ case works analogously. We have for $A(1^+,2^+,3^-)$

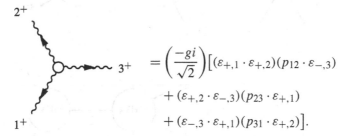

$$= \left(\frac{-gi}{\sqrt{2}}\right)\Big[(\varepsilon_{+,1}\cdot\varepsilon_{+,2})(p_{12}\cdot\varepsilon_{-,3})$$
$$+ (\varepsilon_{+,2}\cdot\varepsilon_{-,3})(p_{23}\cdot\varepsilon_{+,1})$$
$$+ (\varepsilon_{-,3}\cdot\varepsilon_{+,1})(p_{31}\cdot\varepsilon_{+,2})\Big].$$

Now we note for the gauge $\mu_1 = \mu_2 = \mu_3 = \mu$ that

$$\varepsilon_{+,1}\cdot\varepsilon_{+,2} = 0, \qquad \varepsilon_{+,2}\cdot\varepsilon_{-,3} = -\frac{\langle\mu3\rangle[\mu2]}{\langle2\mu\rangle[3\mu]}, \qquad p_{23}\cdot\varepsilon_{+,1} = \sqrt{2}\frac{[12]\langle\mu2\rangle}{\langle1\mu\rangle}.$$

Inserting these into the above amplitude again droping the overall factor $-ig$ yields

$$A\left(1^+,2^+,3^-\right) = -\frac{\langle\mu3\rangle^2}{\langle1\mu\rangle\langle2\mu\rangle}[12].$$

The same argument as before lets us set $\lambda_2 = a\lambda_1$ and $\lambda_3 = b\lambda_1$ and from momentum conservation $\lambda_1(\tilde{\lambda}_1 + a\tilde{\lambda}_2 + b\tilde{\lambda}_3) = 0$ we deduce

$$a = \frac{[31]}{[23]}, \qquad b = \frac{[12]}{[23]}.$$

Hence we have shown that

$$A(1^+, 2^+, 3^-) = -\frac{[12]^3}{[23][31]},$$

as claimed.

Exercise 2.2

We want to determine the NMHV gluon amplitude $A_6^{\text{tree}}(1^+, 2^+, 3^+, 4^-, 5^-, 6^-)$. A shift in 1^+ and 6^- leads to the BCFW recursion relation

$$A_6^{\text{tree}}(1^+, \ldots, 6^-)$$
$$= \sum_{i,s} A_i\left(\hat{1}, 2, \ldots, i-1, -\hat{P}_i^s(z_{P_i})\right)\frac{1}{P_i^2}A_{n+2-i}\left(\hat{P}_i^s(z_{P_i}), i, \ldots, n-1, \hat{n}\right).$$

Diagramatically we have two on-shell diagrams contributing

which we denote by (1) and (2). Using the minus sign convention $|-P] = -|P]$ and $|-P\rangle = |P\rangle$ we have for the first diagram (1)

$$(1) = \frac{[\hat{1}2]^3}{[2\hat{P}_{12}][\hat{P}_{12}\hat{1}]} \times \frac{1}{\langle 12\rangle[21]} \times \frac{[\hat{P}_{21}3]}{[34][45][5\hat{6}][\hat{6}\hat{P}_{12}]}. \qquad (B.10)$$

Writing $P_{ij} = P_i + P_j$ we have $z_P = \frac{P_{12}^2}{\langle 6|P_{12}|1]} = \frac{\langle 12\rangle[21]}{\langle 62\rangle[21]} = \frac{\langle 12\rangle}{\langle 62\rangle}$ and hence

$$|\hat{1}] = |\hat{1}], \qquad |\hat{1}\rangle = |1\rangle - \frac{\langle 12\rangle}{\langle 62\rangle}|6\rangle, \qquad |\hat{6}\rangle = |6\rangle, \qquad |\hat{6}] = |6] + \frac{\langle 12\rangle}{\langle 62\rangle}|1].$$

(B.11)

Furthermore one has

$$\hat{P}_{12} = \hat{P}_1 + P_2 = \lambda_2\tilde{\lambda}_2 + \tilde{\lambda}_1\left(\lambda_1 - \frac{\langle 12\rangle}{\langle 62\rangle}\lambda_6\right)$$

$$= \lambda_2\tilde{\lambda}_2 + \langle 62\rangle^{-1}\left(\langle 62\rangle\lambda_1 + \langle 21\rangle\lambda_6\right) = \lambda_2\left(\tilde{\lambda}_2 + \frac{\langle 61\rangle}{\langle 62\rangle}\tilde{\lambda}_1\right),$$

where we used the Fierz identity in the last step. Hence $|\hat{P}_{12}\rangle = |2\rangle$ and $|\hat{P}_{12}] = |2] + \frac{\langle 61\rangle}{\langle 62\rangle}|1]$. Combining the above we deduce (again using $P_{ij} = P_i + P_j$)

$$[2\hat{P}_{12}] = \frac{\langle 61\rangle}{\langle 62\rangle}[21], \qquad [\hat{P}_{12}\hat{1}] = [21], \qquad [5\hat{6}] = \frac{\langle 5|P_{16}|2]}{62},$$

$$[\hat{P}_{12}3] = \frac{\langle 6|P_{12}|3]}{\langle 62\rangle}, \qquad [\hat{6}\hat{P}_{12}] = -\frac{P_{26}^2 + P_{12}^2 + P_{16}^2}{\langle 62\rangle}.$$

Then we find for the on-shell diagram (1) of Eq. (B.10) the total contribution

$$(1) = \frac{\langle 6|P_{12}|3]^3}{\langle 61\rangle\langle 12\rangle[34][45][5|P_{16}|2\rangle} \frac{1}{P_{26}^2 + P_{12}^2 + P_{16}^2}.$$

(B.12)

Moving on to the second contribution (2) we are led to consider

$$(2) = \frac{\langle 4\hat{P}_{56}\rangle^3}{\langle \hat{P}_{56}\hat{1}\rangle\langle \hat{1}2\rangle\langle 23\rangle\langle 34\rangle} \times \frac{1}{\langle 56\rangle[65]} \times \frac{\langle 5\hat{6}\rangle^3}{\langle 6\hat{P}_{56}\rangle\langle \hat{P}_{56}5\rangle}.$$

(B.13)

Now the shift parameter z_P takes the value $z_P = \frac{[65]}{[51]}$ and we may deduce in analogy to the considerations above that

$$|\hat{1}] = |1], \qquad |\hat{1}\rangle = |1\rangle + \frac{[56]}{[51]}|6\rangle, \qquad |\hat{P}_{56}\rangle = |5\rangle + \frac{[16]}{[15]}|6\rangle.$$

This entails the relations

$$\langle 4\hat{P}_{56}\rangle = \frac{\langle 4|P_{56}|1]}{[51]}, \qquad \langle \hat{P}_{56}\hat{1}\rangle = \frac{P_{15}^2 + P_{56}^2 + P_{16}^2}{[15]}, \qquad \langle 5\hat{6}\rangle = \langle 56\rangle,$$

$$\langle 6\hat{P}_{56}\rangle = \langle 65\rangle, \qquad \langle \hat{P}_{56}5\rangle = \frac{[16]}{[15]}\langle 65\rangle, \qquad \langle \hat{1}2\rangle = \frac{[5|P_{16}|2\rangle}{[51]}.$$

Plugging these into (2) and simplifying terms we arrive at

$$(2) = \frac{\langle 4|P_{56}|1]^3}{\langle 23\rangle\langle 34\rangle[16][65][5|P_{16}|2\rangle} \frac{1}{P_{15}^2 + P_{56}^2 + P_{16}^2}. \tag{B.14}$$

Combining the contributions (1) and (2) we finally conclude

$$A_6^{\text{tree}}(1^+, 2^+, 3^+, 4^-, 5^-, 6^-)$$

$$= \frac{\langle 6|P_{12}|3]^3}{\langle 61\rangle\langle 12\rangle[34][45][5|P_{16}|2\rangle} \frac{1}{P_{26}^2 + P_{12}^2 + P_{16}^2}$$

$$+ \frac{\langle 4|P_{56}|1]^3}{\langle 23\rangle\langle 34\rangle[16][65][5|P_{16}|2\rangle} \frac{1}{P_{15}^2 + P_{56}^2 + P_{16}^2}. \tag{B.15}$$

Exercise 2.3

As suggested we consider the collinear limit in the $(++)$-channel with $\lambda_5 = \sqrt{z}\lambda_P$ and $\lambda_6 = \sqrt{1-z}\lambda_P$

$$A_6^{\text{tree}}(1^-, 2^-, 3^+, 4^+, 5^+, 6^+)$$

$$\xrightarrow{5\|6} \frac{1}{\sqrt{z(1-z)}\langle 56\rangle} \frac{i\langle 12\rangle^4}{\langle 12\rangle\langle 23\rangle\langle 34\rangle\langle 4P\rangle\langle P1\rangle}$$

$$\overset{!}{=} \text{Split}_-^{\text{tree}}(z, 5^+, 6^+) A_5^{\text{tree}}(1^-, 2^-, 3^+, 4^+, P^+)$$

$$+ \text{Split}_+^{\text{tree}}(z, 5^+, 6^+) A_5^{\text{tree}}(1^-, 2^-, 3^+, 4^+, P^-).$$

By comparing the limiting expression on the top line to the terms of the lower lines we see that the last term is absent in the limiting expression. As

$$A_5^{\text{tree}}(1^-, 2^-, 3^+, 4^+, P^-) \neq 0,$$

we deduce that

$$\text{Split}_+^{\text{tree}}(z, a^+, b^+) = 0,$$

in agreement with Eq. (2.41).

Exercise 2.4

Taking the leg 5^- of Eq. (2.58) to the soft limit we have the reduced total momentum conservation condition $p_1 + p_2 + p_3 + p_4 + p_6 = 0$. We pull out the pole term $([45][56])^{-1}$ and find in the limit $p_5 \to 0$ using $p_{26}^2 + p_{12}^2 + p_{16}^2 = p_{34}^2$

$$A_6^{\text{tree}}(1^+, 2^+, 3^+, 4^-, 5^-, 6^-)$$

$$\xrightarrow{5^- \to 0} \frac{1}{[5|p_{16}|2\rangle[45][56]} \left(\frac{\langle 6| \overset{-p_4}{\widehat{p_{12}}} |3]^3[56]}{\langle 61\rangle\langle 12\rangle[34]\langle 34\rangle[43]} + \frac{\langle 46\rangle^3[61]^3[54]}{\langle 23\rangle\langle 34\rangle[16]\langle 16\rangle[61]} \right)$$

$$= \frac{1}{[5|p_{16}|2)[45][56]} \frac{\langle 46 \rangle^3}{\langle 12 \rangle \langle 23 \rangle \langle 34 \rangle \langle 61 \rangle} \left([34][56]\langle 23 \rangle + [61][54]\langle 13 \rangle \right).$$

The two terms in the bracket may be simplified as follows

$$(\cdots) = \underbrace{[56][43]}_{-[54][36]-[53][64]} \langle 32 \rangle + [54][61]\langle 12 \rangle = [54][6|\underbrace{p_{13}}_{-p_4}|2 \rangle + [53][46]\langle 32 \rangle$$

$$= [46]\left([54]\langle 42 \rangle + [53]\langle 32 \rangle \right) = [46][5|p_{34}|2 \rangle = -[46][5|p_{16}|2 \rangle.$$

Plugging this into the above we indeed find

$$A_6^{\text{tree}}\left(1^+, 2^+, 3^+, 4^-, 5^-, 6^- \right) \xrightarrow{5^- \to 0} -\frac{[46]}{[45][56]} \frac{\langle 46 \rangle^3}{\langle 12 \rangle \langle 23 \rangle \langle 34 \rangle \langle 61 \rangle}$$

$$= \text{Soft}\left(4, 5^-, 6 \right) A_5^{\text{tree}}\left(1^+, 2^+, 3^+, 4^-, 6^- \right),$$

which is the expected result consistent with factorization. As both terms in the result contribute.

Exercise 2.5

Let us consider the collinear limit $3^- \parallel 4^+$ with

$$\lambda_3 \to \sqrt{z} \lambda_P, \qquad \lambda_4 \to \sqrt{1-z} \lambda_P$$

of the quark-gluon amplitude $A_5^{\text{tree}}(1_{\bar{q}}^-, 2_q^+, 3^-, 4^+, 5^+)$ to wit

$$A_5^{\text{tree}}\left(1_{\bar{q}}^-, 2_q^+, 3^-, 4^+, 5^+ \right) \xrightarrow{3\|4} \text{Split}_+^{\text{tree}}\left(z, 3^-, 4^+ \right) A_4^{\text{tree}}\left(1_{\bar{q}}^-, 2_q^+, P^-, 5^+ \right)$$

$$+ \text{Split}_-^{\text{tree}}\left(z, 3^-, 4^+ \right) \underbrace{A_4^{\text{tree}}\left(1_{\bar{q}}^-, 2_q^+, P^+, 5^+ \right)}_{=0}$$

$$= \frac{z^2}{\sqrt{z(1-z)}\langle 34 \rangle} \frac{\langle 1P \rangle^3 \langle 2P \rangle}{\langle 12 \rangle \langle 2P \rangle \langle P5 \rangle \langle 51 \rangle}, \tag{B.16}$$

where the 4-point amplitude of Eq. (2.62) was inserted. The limiting form of Eq. (B.16) suggests the original amplitude before the limit to take the form

$$A_5^{\text{tree}}\left(1_{\bar{q}}^-, 2_q^+, 3^-, 4^+, 5^+ \right) = \frac{\langle 13 \rangle^3 \langle 23 \rangle}{\langle 12 \rangle \langle 23 \rangle \langle 34 \rangle \langle 45 \rangle \langle 51 \rangle}.$$

In fact the form above lets one conjecture the multiplicity n form

$$A_n^{\text{tree}}\left(1_{\bar{q}}^-, 2_q^+, 3^-, \ldots, n^+ \right) = \frac{\langle 13 \rangle^3 \langle 23 \rangle}{\langle 12 \rangle \langle 23 \rangle \langle 34 \rangle \cdots \langle n1 \rangle}. \tag{B.17}$$

By analogy to Eq. (B.16) one easily convinces oneself that the conjectured form of the n-point amplitude Eq. (B.17) is consistent with the collinear $3^- \parallel 4^+$ and $i^+ \parallel (i+1)^+$ for $i = 4, \ldots, n-1$ limits. Let us also study two soft limits of Eq. (B.17).

First we take $\lambda_3 \to 0$. Then we immediately see that

$$A_n^{\text{tree}}\left(1_{\bar{q}}^-, 2_q^+, 3^-, 4^+, \ldots, n^+\right)$$

$$\xrightarrow{3^- \to 0} 0 \stackrel{!}{=} \text{Soft}^{\text{tree}}\left(2, 3^-, 4\right) A_n^{\text{tree}}\left(1_{\bar{q}}^-, 2_q^+, 4^+, \ldots, n^+\right).$$

This results implies that

$$A_n^{\text{tree}}\left(1_{\bar{q}}^-, 2_q^+, 3^+, \ldots, n^+\right) = 0, \tag{B.18}$$

being consistent with Eq. (2.63). Taking the soft limit $4^+ \to 0$ (or any other positive helicity gluon leg) on the other hand again checks the consistency of Eq. (B.17)

$$A_n^{\text{tree}}\left(1_{\bar{q}}^-, 2_q^+, 3^-, 4^+, \ldots, n^+\right) \xrightarrow{4^+ \to 0} \underbrace{\frac{\langle 35 \rangle}{\langle 34 \rangle \langle 45 \rangle}}_{\text{Soft}^{\text{tree}}(3, 4^+, 5)} A_{n-1}^{\text{tree}}\left(1_{\bar{q}}^-, 2_q^+, 3^-, 5^+, \ldots, n^+\right).$$

Exercise 2.6

Using the on-shell recursion and the form of the three-point scalar-gluon amplitudes of Sect. 2.5 one has

$$A_4\left(1^+, 2_\phi, 3_{\bar{\phi}}, 4^-\right) = A_3\left(\hat{1}^+, 2_\phi, -\hat{P}_{\bar{\phi}}\right) \frac{1}{P^2 - m^2} A_3\left(\hat{P}_\phi, 3_{\bar{\phi}}, 4^-\right)$$

$$= -\frac{\langle q_1 | \hat{P} | \hat{1}]}{\langle q \hat{1} \rangle} \frac{1}{P^2 - m^2} \frac{\langle \hat{4} | p_3 | q_2]}{[\hat{4} q_2]}.$$

With the gauge choice $q_1 = \hat{p}_4$ and $q_2 = \hat{p}_1$ along with the identities $|\hat{4}\rangle = |4\rangle$ and $|\hat{1}] = |1]$ one has

$$\langle q_1 \hat{1} \rangle = 0, \qquad [\hat{4} \hat{1}] = [\hat{4} 1] = [41],$$

$$\langle q_1 | \hat{P} | \hat{1}] = \langle \hat{4} | \hat{P} | 1] = -\langle 4 | p_3 + \hat{p}_4 | 1] = -\langle 4 | p_3 | 1],$$

$$\langle \hat{4} | p_3 | q_2] = \langle \hat{4} | p_3 | \hat{1}] = \langle 4 | p_3 | 1].$$

Plugging these into the above yields the final compact result

$$A_4\left(1^+, 2_\phi, 3_{\bar{\phi}}, 4^-\right) = -\frac{\langle 4 | p_3 | 1]^2}{\langle 41 \rangle [14][(p_1 + p_2)^2 - m^2]}.$$

Exercise 2.7

It suffices to consider only the single-particle representation quoted in Eq. (2.105). The commutation relations with d

$$\left[d, p^{\alpha\dot{\alpha}}\right] = p^{\alpha\dot{\alpha}}, \qquad [d, k_{\alpha\dot{\alpha}}] = k_{\alpha\dot{\alpha}}, \qquad [d, m_{\alpha\beta}] = 0 = [d, \overline{m}_{\dot{\alpha}\dot{\beta}}],$$

are manifest from simple counting by noting the commutators $[d, \lambda^\alpha] = +\lambda^\alpha$ and $[d, \partial_\alpha] = -\partial_\alpha$. It remains to compute the commutator $[k_{\alpha\dot\alpha}, p^{\beta\dot\beta}]$. One easily establishes

$$\left[k_{\alpha\dot\alpha}, p^{\beta\dot\beta}\right] = \left[\partial_\alpha, \lambda^\beta\tilde\lambda^{\dot\beta}\right]\partial_{\dot\alpha} + \partial_\alpha\left[\partial_{\dot\alpha}, \lambda^\beta\tilde\lambda^{\dot\beta}\right] = \delta_\alpha^\beta\tilde\lambda^{\dot\beta}\partial_{\dot\alpha} + \delta_{\dot\alpha}^{\dot\beta}\lambda^\beta\partial_\alpha + \delta_\alpha^\beta\delta_{\dot\alpha}^{\dot\beta}.$$

Using Eq. (2.98) with raised index

$$\varepsilon^{\alpha\Gamma}\lambda_\Gamma\partial_\beta = \varepsilon^{\alpha\Gamma}\lambda_{(\alpha}\partial_{\beta)} + \frac{1}{2}\underbrace{\varepsilon^{\alpha\Gamma}\varepsilon_{\Gamma\beta}}_{=\delta_\beta^\alpha}\lambda^\Gamma\partial_\Gamma,$$

and the sister equation with dotted indices one easily concludes

$$\left[k_{\alpha\dot\alpha}, p^{\beta\dot\beta}\right] = \delta_\alpha^\beta\varepsilon^{\dot\beta\dot\Gamma}\overline{m}_{\dot\alpha\dot\Gamma} + \delta_{\dot\alpha}^{\dot\beta}\varepsilon^{\beta\Gamma}m_{\alpha\Gamma} + \delta_\alpha^\beta\delta_{\dot\alpha}^{\dot\beta}\left(\frac{1}{2}\lambda^\Gamma\partial_\Gamma + \frac{1}{2}\tilde\lambda^{\dot\Gamma}\partial_{\dot\Gamma} + 1\right),$$

which proves Eq. (2.103).

Exercise 2.8

Using the inversion transformation $I \cdot x^\mu = \frac{x^\mu}{x^2}$ and the translation transformation $P^\mu \cdot x^\mu = x^\mu - a^\mu$ we have with $K^\mu = I \cdot P^\mu \cdot I$

$$K^\mu \cdot x^\mu = I \cdot P^\mu \cdot \frac{x^\mu}{x^2} = I \cdot \frac{x^\mu - a^\mu}{(x-a)^2} = \frac{\frac{x^\mu}{x^2} - a^\mu}{(\frac{x^\mu}{x^2} - a^\mu)^2} = \frac{x^\mu - a^\mu}{(x^\mu - a^\mu x^2)^2}x^2$$

$$= \frac{x^\mu - a^\mu x^2}{1 - 2a \cdot x + a^2 x^2},$$

as claimed.

Exercise 2.9

(a) We start with $\delta^{(2)}(\lambda^\alpha a + \mu^a b)$ for Grassmann even a and b. Then we have

$$\delta^{(2)}(\lambda^\alpha a + \mu^a b) = \delta(\lambda^1 a + \mu^1 b)\delta(\lambda^2 a + \mu^2 b) = \frac{\delta(a + \frac{\mu^1}{\lambda^1}b)}{|\lambda^1|}\delta\left(b\left(\mu^2 - \frac{\mu^1\lambda^2}{\lambda^1}\right)\right)$$

$$= \frac{\delta(a)\delta(b)}{|\lambda^1||\mu^2 - \frac{\mu^1\lambda^2}{\lambda^1}|} = \frac{\delta(a)\delta(b)}{|\langle\lambda\mu\rangle|}, \tag{B.19}$$

where we used $|\lambda^1\mu^2 - \mu^1\lambda^2| = |\langle\lambda\mu\rangle|$ in the last step.

(b) In the fermionic case one arrives at the result by simple multiplication

$$\delta^{(2)}(\lambda^\alpha a + \mu^a b) = (\lambda^1 a + \mu^1 b)(\lambda^2 a + \mu^2 b) = ab(\lambda^1\mu^2 - \lambda^2\mu^1) = \delta(a)\delta(b)\langle\lambda\mu\rangle.$$

(c) Finally we turn to the evaluation of the fermionic delta-function Eq. (2.153). For the three-point case we have

$$\delta^{(8)}(q^{\alpha A}) = \prod_{A=1}^{4} \delta^{(2)}(\lambda_1^\alpha \eta_1^A + \lambda_2^\alpha \eta_2^A + \lambda_3^\alpha \eta_3^A). \tag{B.20}$$

The spinor λ_3^α may be expressed in the basis of the non-collinearly assumed λ_1^α and λ_2^α:

$$\lambda_3^\alpha = x_1 \lambda_1^\alpha + x_2 \lambda_2^\alpha \quad \Rightarrow \quad \langle 13 \rangle = x_2 \langle 12 \rangle, \quad \langle 23 \rangle = x_1 \langle 21 \rangle.$$

Inserting this into Eq. (B.20) we obtain

$$\delta^{(8)}(q^{\alpha A}) = \prod_{A=1}^{4} \delta^{(2)}\left(\lambda_1^\alpha\left(\eta_1^A + \frac{\langle 23 \rangle}{\langle 21 \rangle}\eta_3^A\right) + \lambda_2^\alpha\left(\eta_2^A + \frac{\langle 31 \rangle}{\langle 21 \rangle}\eta_3^A\right)\right)$$

$$= \langle 12 \rangle^4 \delta^{(4)}\left(\eta_1^A + \frac{\langle 23 \rangle}{\langle 21 \rangle}\eta_3^A\right)\delta^{(4)}\left(\eta_2^A + \frac{\langle 31 \rangle}{\langle 21 \rangle}\eta_3^A\right),$$

where we have used Eq. (B.19) in the last step.

Exercise 2.10

The generalization of the discussion in Sect. 2.7.6 to the n-point case proceed in great analogy. We start with the recursion formula of Eq. (2.144) for $p = 0$

$$\mathbb{A}_n^{MHV} = \int \frac{d^4 \eta_P}{P^2} \mathbb{A}_3^{\overline{MHV}}(z_P) \mathbb{A}_{n-1}^{MHV}(z_P)$$

$$= -\int \frac{d^4 \eta_P}{P^2}$$

$$\times \frac{\delta^{(4)}(\eta_1[2,\hat{P}] + \eta_2[\hat{P}1] + \eta_P[12])\delta^{(8)}(\lambda_{\hat{P}}\eta_P + \lambda_3\eta_3 + \cdots + \lambda_n\hat{\eta}_n)}{[12][2\hat{P}][\hat{P}1]\langle \hat{P}3 \rangle \langle 34 \rangle \cdots \langle n-1n \rangle \langle n\hat{P} \rangle}, \tag{B.21}$$

where we assume the super-MHV formula to hold for $(n-1)$-points. Localizing η_P through the fermionic $\delta^{(4)}$-function as in Eq. (2.159) and inserting this into the remaining fermioniv $\delta^{(8)}$ function yields the total supermomentum conservation in analogy to Eq. (2.160)

$$\delta^{(8)}\left(-\frac{\lambda_{\hat{P}}}{[12]}(\eta_1[2\hat{P}] + \eta_2[\hat{P}1]) + \lambda_3\eta_3 + \cdots + \lambda_n\hat{\eta}_n\right)$$

$$= \delta^{(8)}\left(-\frac{\hat{\lambda}_1\eta_1[21] + \hat{\lambda}_2\eta_2[21]}{[12]} + \lambda_3\eta_3 + \cdots + \lambda_n\hat{\eta}_n\right) = \delta^{(8)}(q).$$

Then all that remains is the consideration of the bosonic factors in Eq. (B.21). We have with $\hat{P} = \hat{p}_1 + p_2$

$$-\frac{1}{P^2}[12]^4 \frac{1}{[12][2\hat{P}]\underbrace{[\hat{P}1]\langle\hat{P}3\rangle}_{=[21]\langle23\rangle}\langle34\rangle\cdots\langle n-1n\rangle\langle n\hat{P}\rangle}$$

$$\underline{\underline{[2\hat{P}]\langle n\hat{P}\rangle=[21]\langle n\hat{1}\rangle=[21]\langle n1\rangle}} \qquad \frac{1}{\langle n1\rangle\langle12\rangle\langle23\rangle\langle34\rangle\cdots\langle n-1n\rangle},$$

which completes the proof.

Exercise 3.1

(a) We begin with the computation of d_{23}. Using the labeling of Eq. (3.157) the three on-shell constraints for l_4, l_3 and l_5 are

$$l_4^2 = 0, \qquad l_3^2 = (l_4 - p_4)^2 = 0, \qquad l_5^2 = (l_4 + p_5)^2 = 0$$

in combination with the MHV$_3$ and $\overline{\text{MHV}}_3$ to the left and right of l_4 stateing that $\lambda_{l_4} \sim \lambda_4$ and $\tilde{\lambda}_{l_4} \sim \tilde{\lambda}_5$ we then find the solution to the above three conditions in the form

$$l_4^{\alpha\dot\alpha} = \xi\lambda_4^\alpha\tilde{\lambda}_5^{\dot\alpha}.$$

The remaining constraint $l_1^2 = (l_4 + p_1 + p_5)^2$ then fixes the constant ξ to the value $\xi = \langle15\rangle/\langle41\rangle$. Again there is only one consistent solution to the quadruple cut conditions due to the three-point kinematics. For the coefficient d_{23} we have the product of the MHV$_4$ amplitude with two $\overline{\text{MHV}}_3$ and one MHV$_3$ amplitudes

$$d_{23} = \frac{1}{2}\frac{\langle l_12\rangle^3}{\langle23\rangle\langle3l_3\rangle\langle l_3l_1\rangle}\frac{[4l_4]^3}{[l_4l_3][l_34]}\frac{\langle l_5l_4\rangle^3}{\langle l_45\rangle\langle5l_5\rangle}\frac{[l_1l_5]^3}{[l_51][1l_1]}$$

$$= \frac{1}{2}\frac{[4|l_4l_5l_1|2\rangle^3}{\langle23\rangle\langle5|l_5|1]\langle3|l_3|4]\langle5|l_4l_3l_1|1]}. \tag{B.22}$$

Using the relations $l_3 = l_4 - p_4$, $l_5 = l_4 + p_5$ and $l_1 = l_4 + p_1 + p_5$ one computes

$$[4|l_4l_5l_1|2\rangle = -s_{51}s_{45}\frac{\langle12\rangle}{\langle41\rangle}, \qquad \langle5|l_5|1] = s_{51}\frac{\langle54\rangle}{\langle41\rangle},$$

$$\langle3|l_3|4] = \frac{\langle15\rangle}{\langle41\rangle}\langle34\rangle[54], \qquad \langle5|l_4l_3l_1|1] = s_{51}s_{45}\frac{\langle45\rangle}{\langle41\rangle}.$$

Plugging these results into Eq. (B.22) yields

$$d_{23} = -\frac{1}{2}s_{51}s_{45}\frac{\langle12\rangle^3}{\langle23\rangle\langle34\rangle\langle45\rangle\langle51\rangle} = \frac{i}{2}s_{51}s_{45}A_5^{\text{tree}}\left(1^-, 2^-, 3^+, 4^+, 5^+\right),$$

as claimed in Eq. (3.157).

(b) For the evaluation of d_{34} using the labeling of Eq. (3.157) the three on-shell constraints for l_1, l_5 and l_2 read

$$l_1^2 = 0, \qquad l_5^2 = (l_1 - p_1)^2 = 0, \qquad l_2^2 = (l_1 + p_2)^2 = 0,$$

together with the three-vertex kinematics based conditions $\lambda_{l_1} \sim \lambda_2$ and $\tilde{\lambda}_{l_1} \sim \tilde{\lambda}_1$ implies the ansatz for l_1

$$l_1^{\alpha\dot\alpha} = \xi \lambda_2^\alpha \tilde{\lambda}_1^{\dot\alpha}.$$

The remaining condition $l_4^2 = (l_1 - p_1 - p_5)^2 = 0$ is obeyed for $\xi = \langle 15 \rangle / \langle 25 \rangle$. The coefficient d_{34} then follows from the product of the four tree-level amplitudes depicted in Eq. (3.157)

$$
\begin{aligned}
d_{34} &= \frac{1}{2} \frac{\langle l_4 l_2 \rangle^3}{\langle l_2 3 \rangle \langle 34 \rangle \langle 4 l_4 \rangle} \frac{[l_4 5]^3}{[5 l_5][l_5 l_4]} \frac{\langle 1 l_1 \rangle^3}{\langle l_1 l_5 \rangle \langle l_5 1 \rangle} \frac{[l_2 l_1]^3}{[l_1 2][2 l_2]} \\
&= \frac{1}{2} \frac{\langle 1 | l_1 l_2 l_4 | 5]^3}{\langle 34 \rangle \langle 4 | l_4 l_5 l_1 | 2] \langle 3 | l_2 | 2] \langle 1 | l_5 | 5]}.
\end{aligned}
\tag{B.23}
$$

Evaluating this using $l_5 = l_1 - p_1$, $l_2 = l_1 + p_2$ and $l_4 = l_1 - p_1 - p_5$ one has

$$\langle 1 | l_1 l_2 l_4 | 5] = s_{12} s_{51} \frac{\langle 12 \rangle}{\langle 25 \rangle}, \qquad \langle 4 | l_4 l_5 l_1 | 2] = s_{12} s_{51} \frac{\langle 45 \rangle}{\langle 25 \rangle},$$

$$\langle 3 | l_2 | 2] = \frac{\langle 15 \rangle}{\langle 25 \rangle} \langle 32 \rangle [12], \qquad \langle 1 | l_5 | 5] = -\frac{\langle 12 \rangle}{\langle 25 \rangle} s_{51}.$$

Plugging this into Eq. (B.23) one finds

$$d_{34} = -\frac{1}{2} s_{12} s_{51} \frac{\langle 12 \rangle^3}{\langle 23 \rangle \langle 34 \rangle \langle 45 \rangle \langle 51 \rangle} = \frac{i}{2} s_{12} s_{51} A_5^{\text{tree}} (1^-, 2^-, 3^+, 4^+, 5^+).$$

This proves the result quoted in Eq. (3.157).

Exercise 3.2

(a) $F_{4,m}$ can only depend on the Lorentz invariants s, t, and m^2. Moreover, it has mass dimension zero, therefore it can only depend on dimensionless variables, hence

$$F_{4,m} = F_{4,m}\left(-m^2/s, -m^2/t\right).$$

(b) Switching to dual coordinates, this becomes

$$F_{4,m} = x_{13}^2 x_{24}^2 \int \frac{d^4 x_0}{i\pi^2} \prod_{j=1}^n \frac{1}{-x_{0j}^2 + m^2}.$$

Equation (3.199) gives the following Feynman parametrization

$$F_{4,m} = x_{13}^2 x_{24}^2 \int_0^\infty \prod_{i=1}^4 d\alpha_i \frac{\delta(c_i\alpha_i - 1)}{[-(\alpha_1\alpha_3 x_{13}^2 + \alpha_2\alpha_4 x_{24}^2) + (\sum_{i=1}^4 \alpha_i)^2 m^2]^2}.$$

We can choose $c_i = 1$ in order to simplify the mass term.

(c) One then sees that two Mellin-Barnes parameters are sufficient to factorize the integrand. Indeed,

$$(a + b + c)^{-\lambda} = \frac{1}{\Gamma(\lambda)} \int \frac{dz_1 dz_2}{(2\pi i)^2}$$
$$\times \Gamma(-z_1)\Gamma(-z_2)\Gamma(\lambda + z_1 + z_2)a^{z_1}b^{z_2}c^{-\lambda - z_1 - z_2}.$$

Carrying out the α integrals using Eq. (3.202) one finds

$$F_{4,m} = \int \frac{dz_{1,2}}{(2\pi i)^2} \left(\frac{-s}{m^2}\right)^{1+z_1} \left(\frac{-t}{m^2}\right)^{1+z_2}$$
$$\times \Gamma(-z_1)\Gamma(-z_2)\Gamma(2 + z_1 + z_2)\frac{\Gamma^2(1+z_1)\Gamma^2(1+z_2)}{\Gamma(4 + 2z_1 + 2z_2)}.$$

(d) Here the real part of the integration variables is to be chosen such that

$$-1 < \mathbf{Re}(z_i) < 0,$$

which assures that all intermediate steps leading to this expression are well-defined.

(e) We have $F_{4,m} \propto \int dz_i (m^2)^{-2-z_1-z_2}$. In order to take the limit $m \to 0$, one needs to analytically continue in the z_i variables until the exponent of m^2 becomes positive. (The latter terms vanish as $m \to 0$.) This can be done either by hand, or using the Mathematica implementation *MBasymptotics.m*. The result is

$$F_{4,m} = 2\log(-m^2/s)\log(-m^2/t) - \pi^2 + \mathcal{O}(m^2),$$

or, equivalently,

$$F_{4,m} = \log^2\left(\frac{m^2}{-s}\right) + \log^2\left(\frac{m^2}{-t}\right) - \log^2\left(\frac{-s}{-t}\right) - \pi^2 + \mathcal{O}(m^2).$$

Exercise 3.3

(a) Higher orders in ε. This problem is solved in the main text in Sect. 3.8.3.

(b) Massive triangle integral. We start by generalizing the integral to arbitrary powers of the propagators,

$$F_{3,m}(a_1, a_2, a_3)$$

$$:= \int \frac{d^D k}{i\pi^{D/2}}$$

$$\times \frac{1}{(-k^2 + m^2)^{a_2}(-(k + p_1)^2 + m^2)^{a_1}(-(k + p_2)^2 + m^2)^{a_3}}.$$

Moreover, it is convenient to use the notation

$$1^{\pm} F_{3,m}(a_1, a_2, a_3) = F_{3,m}(a_1 \pm 1, a_2, a_3),$$

and similarly for $2^{\pm}$ and $3^{\pm}$. Then, we derive IBP identities $0 = \int \partial_{k^\mu} v^\mu$, with $v^\mu = k^\mu$. (Other equations can be derived for $v^\mu = p_1$ or $v^\mu = p_2^\mu$, but we will not need them here.) Specifying to $a_i = 1$ and $D = 4$, we have

$$0 = [2m^2(1^+ + 2^+ + 3^+) - 1^+2^- + 3^+2^-]1^{\pm} F_{3,m}(1, 1, 1).$$

On the other hand, differentiating w.r.t. m^2, we have

$$m^2 \partial_{m^2} F_{3,m}(1, 1, 1) = -m^2(1^+ + 2^+ + 3^+).$$

Combining these two equations, we find a differential equation for $F_{3,m}$ in terms of the bubble integral $J(2, 1)$ considered in the main text, with $q^2 = s$,

$$m^2 \partial_{m^2} F_{3,m}(1, 1, 1) = -J(2, 1).$$

For dimensional reasons, we have $F_{3,m}(1, 1, 1) = 1/s f_{3,m}(y)$, where we introduced the variable $y = -m^2/s$. Using Eq. (3.233), we find

$$y \partial_y f_{3,m}(y) = -\frac{1}{\sqrt{1+4y}} \log\left(\frac{\sqrt{1+4y} - 1}{\sqrt{1+4y} + 1}\right),$$

with the boundary condition $f_3(\infty) = 0$. Integrating this equation leads us to Eq. (3.235).

(c) Finite propagator integral from the IBP relations for the triangle subintegral it follows that one can rewrite the original integral Q in terms of simple one-loop propagator integrals. Up to a trivial scale dependence, the latter give rise to factors

$$B(a_1, a_2; D) = \frac{\Gamma(a_1 + a_2 - D/2)\Gamma(D/2 - a_2)\Gamma(D/2 - a_1)}{\Gamma(D - a_1 - a_2)\Gamma(a_1)\Gamma(a_2)}.$$

We have

$$Q(1, 1, 1, 1, 1; D)$$

$$= \frac{-2}{D - 4} B(1, 1; D)[B(3 - D/2, 2; D) - B(1, 2; D)] \frac{1}{(-q^2)^{5-D}}.$$

The occurring Γ functions can be expanded in ε, cf. Eq. (3.204). We find

$$Q(1, 1, 1, 1, 1; 4 - 2\varepsilon) = 6\zeta_3 \frac{1}{(-q^2)} + \mathcal{O}(\varepsilon),$$

and higher orders in ε can also be generated without difficulty.

Exercise 4.1

(a) Here we verify that F_4 of Eq. (4.29) satisfies the conformal Ward identity (4.27). We start by showing that

$$K^\mu x_{ab}^2 = 2(x_a^\mu + x_b^\mu) x_{ab}^2, \tag{B.24}$$

and hence

$$K^\mu \log x_{ab}^2 = 2(x_a^\mu + x_b^\mu).$$

From this it follows that

$$K^\mu F_4 = \Gamma_{\text{cusp}}(a)(x_1^\mu + x_3^\mu - x_2^\mu - x_4^\mu) \log\left(\frac{x_{13}^2}{x_{24}^2}\right),$$

in complete agreement with Eq. (4.27).

(b) A particular solution to the Ward identities (4.27) was given in Ref. [1]. For four and five points, the homogeneous solution of the differential equation is a constant, i.e. no conformal invariants can be built from four or five light-like separated points. For more points, one can build conformal cross-ratios, see Eq. (4.30). Using Eq. (B.24) it is easy to see that these are indeed invariant,

$$K^\mu \frac{x_{ij} x_{kl}}{x_{ik}^2 x_{jl}^2} = 0.$$

Another way to see the invariance is to perform inversions. At six points, there are three independent cross-ratios u, v, w, see Eq. (4.31), and therefore the homogeneous solution is an a priori unconstrained function of u, v, w (and the coupling).

References

1. J.M. Drummond, J. Henn, G.P. Korchemsky, E. Sokatchev, On planar gluon amplitudes/Wilson loops duality. Nucl. Phys. B **795**, 52 (2008). arXiv:0709.2368